HK SPORT SCIENCE MONOGRAPH SERIES
Volume 7

Exercise and Intracellular Regulation of Cardiac and Skeletal Muscle

Michael I. Kalinski, PhD
Brooklyn College of the City University of New York

Alexander Ye. Antipenko, PhD
Mount Sinai School of Medicine, City University of New York

Christopher C. Dunbar, PhD
Brooklyn College of the City University of New York

Donald W. Michielli, PhD
Brooklyn College of the City University of New York

Human Kinetics

Library of Congress Cataloging-in-Publication Data

Exercise and intracellular regulation of cardiac and skeletal muscle / Michael I. Kalinski . . . [et al.].

p. cm. -- (HK sport science monograph series, ISSN 0894-4229 ; v. 7)

Includes bibliographical references.

ISBN 0-87322-725-5

1. Muscles--Physiology. 2. Myocardium--Physiology. 3. Exercise--Physiological aspects. 4. Cyclic adenylic acid. 5. Protein kinase. I. Kalinski, Michael, 1943- . II. Series.

[DNLM: 1. Exercise--physiology. 2. Muscles--metabolism. 3. Phosphorylation. 4. Cyclic AMP--metabolism. 5. Cyclic AMP -Dependent Protein Kinases--metabolism. 6. Second Messenger Systems--physiology. WE 103 E955 1995]

QP321.E94 1995

612'.044--dc20

DNLM/DLC 94-30882

for Library of Congress CIP

ISBN: 0-87322-725-5

ISSN: 0894-4229

Sections of Part I are based on ''Membrane Phosphorylation in Vascular and Cardiac Muscles: Comparative Analysis and Functional Role'' by A.Ye. Antipenko, 1991, *Basic and Applied Myology*, **1**, pp. 209-233. Copyright 1991 by Unipress. Adapted by permission.

Acquisitions Editor: Richard A. Washburn, PhD; **Developmental Editor:** Julia Anderson; **Assistant Editors:** Jacqueline Blakley, Ed Giles, and Julie Ohnemus; **Copyeditor:** Joyce Sexton; **Proofreader:** Jim Burns; **Typesetting and Text Layout:** Yvonne Winsor; **Designer:** Keith Blomberg; **Illustrator:** Tom Janowski; **Printer:** United Graphics

Printed in the United States of America 10 9 8 7 6 5 4 3 2 1

Human Kinetics
P.O. Box 5076, Champaign, IL 61825-5076
1-800-747-4457

Canada: Human Kinetics, Box 24040,
Windsor, ON N8Y 4Y9
1-800-465-7301 (in Canada only)

Europe: Human Kinetics,
P.O. Box IW14, Leeds LS16 6TR, England
(44) 532 781708

Australia: Human Kinetics, 2 Ingrid Street,
Clapham 5062, South Australia
(08) 371 3755

New Zealand: Human Kinetics, P.O. Box 105-231,
Auckland 1
(09) 309 2259

Contents

HK Sport Science Monograph Series

The *HK Sport Science Monograph Series* is another endeavor to provide a useful communication channel for recording extensive research programs by sport scientists. Many publishers have discontinued publishing monographs because they are uneconomical. It is my hope that with the cooperation of authors, the use of electronic support systems, and the purchase of these monographs by sport scientists and libraries we can continue this series over the years.

The series will publish original research reports and reviews of literature that are sufficiently extensive not to lend themselves to reporting in available research journals. Subject matter pertinent both to the broad fields of the sport sciences and to physical education are considered appropriate for the monograph series, especially research in

- sport biomechanics,
- sport physiology,
- motor behavior (including motor control and learning, motor development, and adapted physical activity),
- sport psychology,
- sport sociology, and
- sport pedagogy.

Other titles in this series are:

- *Adolescent Growth and Motor Performances: A Longitudinal Study of Belgian Boys*
- *Biological Effects of Physical Activity*
- *Growth and Fitness of Flemish Girls: The Leuven Growth Study*
- *Kinanthropometry in Aquatic Sports: A Study of World Class Athletes*

Authors who wish to publish in the monograph series should submit two copies of the complete manuscript to the publisher. All manuscripts must conform to the current *APA Publication Manual* and be of a length between 120 and 300 double-spaced manuscript pages. The manuscript will be sent to two reviewers who will follow a review process similar to that used for scholarly journals. The decision with regard to the manuscript's acceptability will be based on its judged contribution to knowledge and on economic feasibility. Publications that are accepted, after all required revisions are made, must be submitted to the publisher on computer disk for electronic transfer to typesetting. No royalties will be paid for monographs published in this series.

Authors wishing to submit a manuscript to the monograph series or desiring further information should write to: Human Kinetics, P.O. Box 5076, Champaign, IL 61825-5076 for further details.

Rainer Martens

Preface

Since the discovery in 1957 of the "second messenger" cyclic AMP (cAMP) by the Nobel laureate E. Sutherland, the central role in energy metabolism of cAMP and the enzymes that phosphorylate proteins has become increasingly apparent. Acknowledgment of researchers in this area continues. In 1992 Edwin G. Krebs and Edmond H. Fisher were awarded the Nobel Prize for their work with second messengers and phosphoproteins.

In addition to the numerous well-publicized findings of Western scientists, many discoveries concerning the properties of cAMP-dependent protein kinase of cardiac, skeletal, and smooth muscle have been made in Eastern Europe. The full results of the findings of M. Kurski, T. Kondratiuk, and Z. Vorobets of the Division of Muscle Biochemistry, Research Institute of Biochemistry, Academy of Science of Ukraine have been published only in Slavic languages. A. Antipenko, formerly of St. Petersburg, has also made significant contributions to the knowledge of cAMP metabolism, cAMP-dependent phosphorylation, and the regulation of Ca^{++} transport in cardiac sarcoplasmic reticulum.

The Ukrainian school of exercise biochemistry was established in the 1930s. In its early days W. Belitser of Ukraine produced significant contributions to basic biochemical concept of oxidative phosphorylation. Other leading Ukrainian scientists included D. Ferdman (studying processes of transamination), A. Palladin (metabolism of protein, glycogen, ATP, creatine phosphate, and lactic acid during exercise training), and E. Kozhukhar (enzymatic degradation of carbohydrates and the effects of vitamins on metabolism in athletes during exercise).

Since the 1960s, the Kiev Institute of Physical Culture, the Kiev Institute of Biochemistry, and the Kiev Research Institute of Endocrinology and Metabolism have established new frontiers in exercise biochemistry in the areas of hormonal and intracellular regulation of metabolism. M. Kalinski has directed this work, assisted at various times by A. Osipenko, I. Zemtsova, V. Kotsuruba, M. Kurski, T. Kondratiuk, V. Lishko, V. Kononenko, V. Stefanow, W. Tiutiunnik, and N. Rudnitska. Areas of investigation have included

- metabolism of catecholamines during acute exercise and exercise training;
- effects of catecholamines encapsulated into liposomes on energy metabolism in cardiac and skeletal muscle, liver, and blood during exercise;
- metabolism of intracellular second messenger cAMP;
- effects of activators of adenylate cyclase (AC) and inhibitors of phosphodiesterase (PDE) on cAMP metabolism in muscle tissues during exercise; and
- properties of protein kinase and the role of cAMP-dependent phosphorylation in the regulation of glycolysis and transport of Ca^{++} in sarcoplasmic reticulum (SR) and sarcolemma (SL) of cardiac and skeletal muscle during exercise training.

Some of the information gained by this group during more than 20 years of effort is being presented for the first time in English in this monograph.

Research in cell signaling has grown extraordinarily. Numerous subspecialties such as metabolism of second messengers, receptor characteristics, properties of protein kinases and protein phosphatases, and the factors controlling gene transcription have come into being.

Another subspecialty is the effect of exercise on signal transduction. Increasingly, science is elucidating the subcellular events through which the additional metabolic requirements of physical activity are met. The knowledge base has progressed from studies of substrate utilization to exploration of the complex molecular events associated with physical exertion. Our knowledge of hormonal events has also progressed greatly. The transduction systems that carry information from the cell membrane into the cellular interior and initiate the biochemical events leading to increased lipolysis, glycogenolysis, and calcium transport have begun to be described. In recognition of the importance of phosphorylation in controlling cellular events, the roles of cAMP, calmodulin, phospholipids, and cGMP and the activation of protein kinases are described in this volume. Of all of these messengers, cAMP-dependent phosphorylation appears to be the most important regulator of contractile activity in cardiac and skeletal muscle. Information concerning the effects of physical activity on cAMP is therefore presented.

During the last decade it has become apparent that cAMP, cGMP, calmodulin, and phospholipids initiate a rapid and reversible phosphorylation of intracellular protein substrates. This occurs either by activation of the corresponding protein kinases or by modulation of the activities of phosphoprotein phosphatases and enzymes of second messenger metabolism. Thus the regulation of cardiac contractility is realized mainly through cyclic nucleotide-dependent and Ca^{++}-dependent phosphorylation of target proteins in the SL, SR, and myofibrils of these cells.

Significantly less is known about the regulatory role of second messengers in skeletal muscle. It does appear, however, that the main role in contractile activity regulation in skeletal muscle also belongs to the cAMP system. Therefore, we shall begin by examining the cAMP-dependent regulation of cardiac and skeletal muscle Ca^{++} status.

The function of heart and skeletal muscle depends on the coordination of electrical, mechanical, and metabolic activities. The prime regulator of these activities seems to be intracellular Ca^{++}. This ion is at a low concentration in the cytoplasm of the myocyte as compared with the extracellular medium. A relatively small Ca^{++} influx serves as a signal that leads to the initiation of contraction. Intracellular Ca^{++} concentration is subject to neural and humoral control mechanisms involved in a symbiotic relationship with second messengers.

The process of coupling chemical and electrical signals at the cell surface to intracellular release of Ca^{++} and ultimately to contraction of muscle fibers is termed *excitation-contraction coupling*. This coupling of cell surface signals to intracellular Ca^{++} release proceeds by different mechanisms in cardiac and skeletal muscle cells. However, in each of these kinds of cells, cAMP-regulated voltage-gated Ca^{++} channels in the cell surface membranes play a key role.

More detailed data on biochemical changes are essential for understanding the regulation of muscle contractility under varying conditions including pathophysiological conditions and physical exercise (Antipenko et al., 1985, 1989, 1992; Antipenko, 1985; Antipenko & Lyzlova, 1985; Kalinski et al., 1977, 1980, 1981, 1982, 1983, 1984, 1985, 1989, 1990). The effects of exercise on cAMP and cAMP-dependent protein kinase have only recently been studied in any detail, and much remains to be discovered in this area.

One primary goal of the present undertaking was to provide a basic review of the extracellular and intracellular components that constitute the regulatory systems of the living cell. Included is a description of the transduction between the two types of components, including the role of Ca^{++} and the role of phosphorylation in the regulation of cardiac and skeletal muscle. The second goal was to explore the experimental evidence that deals with the effects of both acute and chronic exercise on the metabolism of cAMP and protein kinase in cardiac and skeletal muscle.

The format of this book has been geared toward exercise biochemists, physiologists, and the graduate level student. Part I is an overview of phosphorylation; Part II describes the effects of exercise on cAMP and cAMP-dependent protein kinase; and Part III summarizes this information.

The American investigators who provided the basis for our understanding of cAMP metabolism during exercise are too numerous to list here. However, we wish to acknowledge the contributions of P. Buckenmeyer, G. Dohm, A. Goldfarb, L. Oscai, W. Palmer, R. Shephard, and W. Winder to this area of exercise biochemistry.

To our knowledge, this is the first book devoted primarily to the effect of exercise on cAMP and cAMP-dependent protein kinase. We hope that this work will in some small way advance progress in the field of exercise biochemistry.

Acknowledgments

We would like to thank Richard Frey, Rick Washburn, and Julia Anderson of Human Kinetics for their hard work and excellent suggestions during the development of this monograph.

We are indebted to the members of the Department of Exercise Biochemistry from the Kiev State Institute of Physical Culture and the Division of Muscle Biochemistry of the Kiev Research Institute of Biochemistry, both located in Ukraine, for effective, longtime collaboration and support. Michael Kalinski was affiliated with both the Kiev State Institute of Physical Culture and the Research Institute of Biochemistry at the time this monograph was written.

We are also indebted to Christian Zauner, PhD, who thoroughly reviewed this manuscript.

Special thanks to Anna Procyk for encouragement and support, and to V. Arlene Pikoulas for assistance.

PART I

Phosphorylation and the Functional Regulation of Cardiac and Skeletal Muscle

Chapter 1

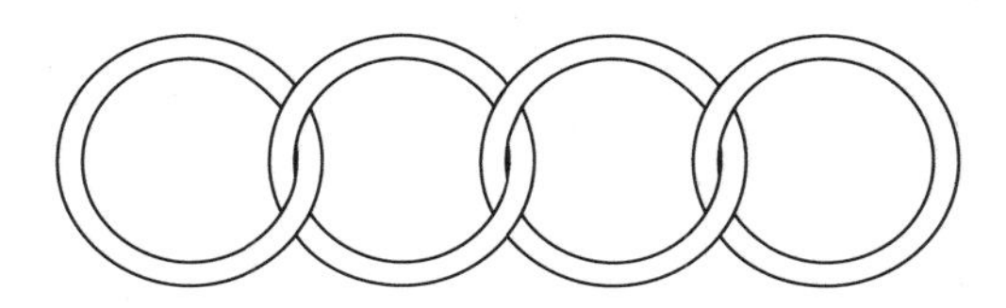

Phosphorylation of Cell Membrane Calcium Channel Proteins

The reaction of protein phosphorylation discussed in chapter 1 may be described by the following formula:

$$\text{Protein-OH} + [\text{gamma }^{32}\text{P}]\ \text{ATP} \xrightarrow{\text{protein kinase}} \text{Protein-O-}^{32}\text{PO}_3\text{H}_2 + \text{ADP}$$

The protein-OH groups, which accept the terminal phosphates of ATP, belong almost exclusively to serine, and occasionally to threonine and tyrosine. Protein phosphoserines undergo considerable ionization at physiological pH values; therefore the phosphorylation or dephosphorylation of serine changes the negative charge of protein molecules. In membranes, alterations of ionic conductivity may be regarded as one of the most significant results of protein phosphorylation.

The phospholipid bilayer of the cell membrane contains protein and glycoprotein molecules, including those participating in Ca^{++} transport. Three ways by which Ca^{++} is transported through the cell membrane can be distinguished as slow Ca^{++} channels, Ca^{++}-ATPase, and Na^{+}-Ca^{++} exchange.

Calcium Channel Proteins in Cardiac Muscle

Myocardial contraction is initiated by the penetration of Ca^{++} into the cell via voltage-dependent Ca^{++} channels of the SL. This amount of Ca^{++}, sometimes called a *trigger*, ranges from 2% to 3% (Wollenberger & Will, 1978) to 20% to 30% (England, 1980) of the amount required in order for a contraction response to manifest itself. Slow Ca^{++} channel blocking and the consequent blocking of Ca^{++} penetration into cardiac muscle cells with antagonists (verapamil, diltiazem, etc.) suppress the contraction completely without exerting any significant influence on the action potential. In other words, the contraction is dissociated from the excitation. These data provide support for the hypothesis (Fabiato & Fabiato,

1979) that Ca^{++} entering the cell stimulates the release of a much greater amount of Ca^{++} from the SR.

It has become known that beta-adrenergic agonists increase the myocardial contractility by opening Ca^{++} channels during depolarization (Akerman, 1982; Reuter, 1983; Sperelakis & Wahler, 1988; Velema et al., 1983). This effect results from the activation of a cAMP-dependent protein kinase (A-kinase) that phosphorylates channel proteins (Velema et al., 1983).

Modulation of Ca^{++} flow through cardiac slow channels can be achieved by injection of cAMP in liposomic envelopes or by pressure-injection cholera toxin (an AC activator) or thermostable A-kinase inhibitor (Sperelakis & Wahler, 1988). The effect of beta-adrenergic agonists can be also reproduced by the injection of methylxanthines (e.g., theophylline), which are cAMP-phosphodiesterase (PDE) inhibitors (Reuter, 1979). All of these findings confirm the hypothesis about Ca^{++} channel phosphorylation.

Fluoride, a phosphatase inhibitor, also renders calcium channels capable of electrodependent activation, probably by hindering their dephosphorylation (Reuter, 1979; Wollenberger & Will, 1978). Hence, a decrease in dephosphorylation rate exerts the same influence on slow Ca^{++} channels as does an acceleration of phosphorylation. It has been shown that protein phosphatase 2B (calcineurin) and phosphatase 2A inhibit the slow Ca^{++} flow through the myocardial SL (Kameyama et al., 1986). However, the leading role in such inhibition seems to belong to protein phosphatase 1, a main phosphatase of the myocardium (Kameyama et al., 1986; McDongall et al., 1991).

Catecholamines, while influencing protein phosphorylation, do not change the parameters of ion permeability of those slow channels in the cardiac SL that are already in action; but they do increase the time during which the channels are open or open new Ca^{++} channels (Sperelakis & Wahler, 1988). Phosphorylation can activate slow Ca^{++} channels by changing their conformation. This either allows the channel to open itself during depolarization or effectively increases the diameter of water-filled pores, providing the possibility for Ca^{++} ions to pass through the channel. Dephosphorylated channels are electrically inactive. Depending on cAMP level and protein phosphatase activity, there is a balance in cardiac cells between the phosphorylated and dephosphorylated channels. Substances that increase myocardial contractility by raising the cAMP level can also increase the proportion of phosphorylated Ca^{++} channels (Lamers, 1985). One of the proteins forming the slow channel itself may serve as a substrate for phosphorylation as well as an adjacent regulatory protein closely connected with the Ca^{++} channel. A similar situation can be found in the complex of phospholamban with Ca^{++}-ATPase in the SR (Sperelakis & Wahler, 1988).

A protein of 11.7 kDa isolated from the cardiac SL and containing a considerable number of Ca^{++}-binding sites was reported to provide a possible substrate for a cAMP-dependent protein kinase of the slow channels (Velema et al., 1983). According to several authors (Jones et al., 1980; Will, 1983), it is the cAMP-dependent phosphorylation of low–molecular-mass proteins of the cardiac SL that may be directly associated with the regulation of slow Ca^{++} channel function.

Velema et al. (1983) described a 17.5 kDa glycoprotein from the cardiac SL that could be phosphorylated by cAMP-dependent protein kinase. This finding is of special interest because glycoproteins take part in the Ca^{++}-ion binding in the glycocalix.

Cardiac SL protein phosphorylation can also be activated by insulin. A proteolipid with molecular mass of 15.5 kDa contains a protein subunit of 3.6 kDa that can be phosphorylated at Ser sites in the presence of insulin (Walaas et al., 1973). Walaas et al. (1973) suggested that this proteolipid could perform a transport function in the membrane while forming the channels permeable for ions and certain water-soluble compounds.

In cardiac SL preparations, a considerable number of protein substrates for A-kinase have been found (Akerman, 1982; Dolphin, 1991; Jones et al., 1980; Reuter, 1979; Velema et al., 1983; Will, 1983). Such a variety of results is often attributable to the differing conditions under which membrane preparation specimens are processed before electrophoresis. For instance, a 24-kDa protein heated at up to 95-100 °C will produce 9-kDa polypeptide fragments; this is not the case when the protein is incubated at 30 °C in the presence of SDS (Lamers & Stinis, 1981). Nonoptimal conditions of pre-phoretic SL membrane treatment can also lead to the appearance of rapidly aggregating high-molecular-mass protein complexes.

cAMP-Dependent Phosphorylation

Among SL proteins, of special interest is calciductin, which was discovered and described by Rinaldi et al. (1981). Calciductin, a protein of 23 kDa, is in the authors' opinion a main target protein for A-kinase in the cardiac SL. Calciductin phosphorylation in the presence of cAMP does not change the properties of Na^{+}-Ca^{++} carriers, but is accompanied by a more than twofold increase in electrodependent Ca^{++} absorption in SL vesicles (Rinaldi et al., 1982). This process probably reflects the functioning of slow Ca^{++} channels. The combination of data obtained led the authors to conclude that beta-agonists realize their regulating action on the myocardium via the phosphorylation of calciductin, which may itself form a channel and/or serve in its phosphorylated form as a Ca^{++} channel activator (Rinaldi, 1981, 1982).

Calciductin phosphorylation can be highly efficient. One reason is that A-kinase is localized in submembrane compartments in close vicinity to calciductin (Jones et al., 1981). In addition, calciductin is an excellent phosphate acceptor. Thus, submaximum calciductin phosphorylation can be achieved with the dissociation of only 6% of the total amount of cAMP-dependent protein kinase holoenzyme (Rinaldi et al., 1982). Hence, calciductin can be efficiently phosphorylated in vivo under minimal cardiac stimulation by beta-agonists.

Calciductin is phosphorylated almost entirely with the participation of A-kinase alone (Sperelakis, 1984). However, the possibility that Ca^{++}-calmodulin-dependent protein kinases (B-kinases) are involved cannot be excluded (Jones et al., 1981). If this is the case, it is possible that the increase in low-molecular-mass substrate phosphorylation that occurs as calcium enters the myocyte may lead to the maintenance of this process for a longer period of time as a result of a positive feedback effect. Indeed, cardiac SL preparations contain endogenous Ca^{++}-dependent protein kinases whose activity will increase to a level several times higher upon addition of calmodulin (Vetter et al., 1982). Maximum Ca^{++} transport stimulation has been observed in the SL in the presence of catalytic A-kinase subunit as well as of calmodulin, the effect being more than additive (Vetter et al., 1982) . Stimulation of Ca^{++} transport through the SL by means of protein phosphorylation takes place under both low and saturating concentrations of this ion, whereas in the SR this is the case only for Ca^{++} concentrations of not more than 1 μmol/L.

Penetration of Ca^{++} into phosphorylated vesicles is inhibited by verapamil, lanthanum, and bivalent cations. Rinaldi et al. (1982) found that these agents do not influence Ca^{++} penetration through the nonphosphorylated channels. Thus, in the authors' opinion, calciductin phosphorylation promotes the opening of new channels that become identical to those already acting in the SL. However, it is noteworthy that these results were obtained during the study of SL preparations probably significantly contaminated with SR vesicles. This suggestion is based on the fact that a high level of oxalate-dependent Ca^{++} absorption (300-400 nmol Ca^{++} per mg protein) was observed in this study. In addition, a significant proportion of the protein studied could have been represented by phospholamban (Manalan & Jones, 1982). Indeed, phospholamban appears in SR preparations at relatively high concentrations of 1.5-2.0 nmol per mg protein (Colyer & Wang, 1991; Tada & Katz, 1982); at the same time, 23-kDa protein contents in SL preparations are about 0.1 nmol per mg (Katz et al., 1983). Therefore, even an insignificant contamination of SL preparations with SR fragments can result in the appearance of additionally assayed amounts of 23-kDa protein. Nevertheless, it is believed that the presence of calciductin in the SL is not caused by contamination with SR fragments and that a protein identical to phospholamban in fact belongs to the SL (Iwasa & Hosey, 1984). Calciductin and phospholamban significantly differ from each other in their capacities for phosphorylation under the influence of various protein kinases (Rinaldi et al., 1982). These differences may be associated both with the specific structure of phosphosubstrates themselves and with different protein and lipid environments in the SL and SR membranes.

Undoubtedly, further investigation using highly purified SL preparations is required in order to evaluate the influence of 23-kDa protein phosphorylation on the Ca^{++} transporting function of the SL. It has even been recommended that the usage of the term *calciductin* with regard to the 23-kDa protein be avoided until the participation of this protein in the regulation of slow Ca^{++} activity is finally proven (Philipson, 1983). Meanwhile, the search for regulatory proteins that can be phosphorylated by cAMP-dependent protein kinase in the slow SL channels is continuing. The search has been stimulated by promising results. Thus, injection of catalytic A-kinase subunits into the cardiac cell is accompanied by an increase in the duration of action potentials and the amplitude of slow Ca^{++} influx at the plateau phase of action potentials. Injection of regulatory subunits of this enzyme into cardiac cells leads to an opposite effect, probably due to the binding of free catalytic subunits (Osterrieder et al., 1982).

Ca^{++} = Dependent Phosphorylation

The data available on the regulatory role of Ca^{++}-calmodulin-dependent phosphorylation in the SL are to some extent contradictory. For instance, as the intracellular Ca^{++} level rises to 10^{-5} mol/L, Ca^{++}-calmodulin-protein kinase complex is being formed and certain protein substrates of the SL, such as 54-kDa and 44-kDa proteins, are being phosphorylated (Tuana et al., 1989). These events are accompanied by the closing of or reduction in the number of Ca^{++} channels, which stops the penetration of Ca^{++} into the cell (Costa et al., 1976). At the same time, it has been demonstrated that calmodulin inhibitors such as trifluoperazine reduce the slow Ca^{++} flow in the heart (Orlov, 1987; Sperelakis & Wahler, 1988). Alpha-adrenergic agonists, in their turn, increase the penetration of Ca^{++} into the cell via slow channels (Sperelakis & Wahler, 1988). Alpha-adrenoreception stimulates the phosphatidylinositol (PI) cycle to generate inositoltrisphosphate (IP_3) and diacylglycerol. The IP_3 promotes Ca^{++} release from the SR depot; diacylglycerol and Ca^{++} activate C-kinase, a Ca^{++}-phospholipid-dependent enzyme (Houslay, 1991). However, there is still no direct evidence for the participation of C-kinase in the regulation of slow Ca^{++} channel permeability, although a considerable number of proteins that can be phosphorylated by this kinase were found in the myocardial SL (Vorobetz et al., 1988).

cGMP = Dependent Phosphorylation

The cyclic nucleotide cGMP may also have important influences on cardiac function. Interestingly, cGMP and cAMP exert opposite effects on the slow Ca^{++} flow in the heart. In frog heart ventricles, cAMP (1-5 μM) stimulates a 10-13-fold increase in slow Ca^{++} influx, whereas cGMP (0.3-20 μM) inhibits this process (Vorobetz et al., 1988). Also, pressure injection of cGMP into cardiac cells rapidly and efficiently inhibits the slow Ca^{++} influx. Similar results have been obtained with the injection of cGMP into liposomic envelopes (Sperelakis & Wahler, 1988). Acetylcholine inhibits slow Ca^{++} flow probably by activating guanylate cyclase associated with muscarinic receptors; this leads to a subsequent increase in cGMP level.

There are three different points of view on the mechanism of inhibitory cGMP action on slow Ca^{++} flow in the heart. Thus, it has been determined that cGMP can inhibit slow Ca^{++} flow only after the rise in cAMP levels in the cardiac cell (Hartzell & Fischmeister, 1986). The authors of this paper believe that inhibitory cGMP action on slow Ca^{++} flow is associated with PDE stimulation, which leads to decreased cAMP levels. To support this hypothesis, the authors adduce the following experimental results: (a) cGMP does not influence slow Ca^{++} flow stimulated by 8-Br-cAMP, PDE-resistant nucleotide analogue, and (b) isobutylmethylxanthine (IBMX), an inhibitor of cGMP-stimulated PDE, reduces the effect of cGMP on slow Ca^{++} flow in the heart (Hartzell & Fischmeister, 1986). Because cGMP-stimulated cAMP PDE has been found in many tissues, cAMP hydrolysis activation can form a universal mechanism underlying cGMP's regulatory action. This hypothesis blends well with the known conclusion on opposite biological effects produced by cAMP and cGMP. The discovery of cGMP-binding proteins differing from G-kinase (a cGMP-dependent enzyme) may indeed testify to the suggestion that cGMP does not always realize its action via cGMP-dependent phosphorylation.

At the same time, there is evidence for another point of view—that it is cGMP-dependent phosphorylation that influences Ca^{++} flow in the heart. The compound 8-Br-cGMP was shown to inhibit Ca^{++} flow without producing decreases in cAMP levels (Sperelakis & Wahler, 1988). This compound also inhibits baseline non-cAMP-stimulated slow Ca^{++} flow in chicken cardiac cells (Wahler et al., 1987). It should be noted that this cGMP analogue, which is capable of penetrating the membrane, does not stimulate PDE but can activate G-kinase. We believe that cGMP inhibits slow Ca^{++} flow directly by G-kinase-dependent phosphorylation of a specific protein. This protein is closely connected with slow Ca^{++} channels in the SL and differs from a protein that can be phosphorylated by A-kinase. The existence of a protein that can be phosphorylated by G-kinase in the myocardial SL has already been reported (Sperelakis & Wahler, 1988). The possibility cannot be excluded that targets for A- or G-kinase action are formed by different amino acid residues within a single protein molecule that forms the channel and/or is closely connected with it.

Finally, the third way in which cGMP exerts its influence on slow Ca^{++} flow in the heart is probably associated with A-kinase activity modulation. Thus, cGMP causes the increase in phosphorylation of the regulatory subunit of A-kinase in guinea pig cardiac SL and the corresponding change in cAMP-dependent protein kinase activity (Cuppoletti et al., 1988). It is believed that cGMP-dependent phosphorylation of the A-kinase regulatory subunit is a universal regulation mechanism of physiological importance (Cuppoletti et al., 1988).

Fast Ca^{++} channels have also been found in heart muscle. These channels can be inactivated much more rapidly than the slow ones (Caroni & Carafoli, 1983). The function of these channels is not yet clear, and cAMP and calcium antagonists exert virtually no effect on their performance (Sperelakis & Wahler, 1988).

The different effects exerted by cAMP and cGMP on Ca^{++} flow in the various tissues can probably be explained by the difference between these tissues in cGMP-dependent PDE activation (Stoclet et al., 1988). Thus, in the myocardium and vascular smooth muscle, two different PDE forms have been found: PDE I, an enzyme that is stimulated by calmodulin and catalyzes the hydrolysis of both cAMP and cGMP; and PDE IV, an enzyme form that is competitively inhibited by cGMP and catalyzes the hydrolysis of cAMP with low K_m (England, 1976). It has been determined that in the heart, hormones and neurotransmitters exert mainly a stimulating influence on PDE activity with low K_m properties. However, physiological inhibitors of this enzyme form also exist (Mery et al., 1990); glucagon is one of these, glucagon receptors being coupled with PDE IV by G proteins. Inhibition of PDE IV seems to cause a positive inotropic glucagon effect in the heart of those animal species, such as guinea pigs, mice, and monkeys, in which this peptide does not stimulate AC activity (Mery et al., 1990).

At the same time PDE II, an enzyme form hydrolyzing both cAMP and cGMP, and activated by cGMP, was found in the heart. This enzyme form cannot be observed in vascular muscle cells. In contrast, G-PDE was discovered in aortic smooth muscle cells. This enzyme form specifically hydrolyzes cGMP but not cAMP (Stoclet et al., 1988) and could not be found in cardiac cells. Thus, the two different PDE isoenzymes in cardiac and vascular smooth muscle cells cause two different modes of regulation for both cyclic nucleotide levels in these cells and certain cyclic nucleotide-dependent processes including the flow of Ca^{++} through the slow channels. In the heart, but not in vascular smooth muscle, cGMP can inhibit cAMP-stimulated Ca^{++} flow through the voltage-dependent channels with PDE II (Hartzell & Fischmeister, 1986). Also, cGMP can inhibit beta-adrenergic activation in the heart via the PDE mechanism (Fischmeister & Hartzell, 1986).

Heart perfusion with compounds that increase the cGMP level leads to a slight negative inotropic effect. The action of cholinergic agonists in the heart seems to be associated with the activation of dephosphorylation (Hartzell & Titus, 1982) rather than with the stimulation of cAMP hydrolysis by cGMP-activated PDE (England, 1980). It should be noted that Hartzell and Titus (1982) did not present any data on the direct influence of cGMP on phosphoprotein phosphatase activity. In their experiments, these authors could determine only that carbamylcholine, a muscarinic cholinergic agonist, inhibited ^{32}P incorporation into a cardiac myofibrillar C protein.

Regulation of slow Ca^{++} flow through the cell membrane of cardiac cells may be associated with the phosphorylation of a 15-kDa protein. In the heart, phosphorylation of this 15-kDa protein is accompanied by a positive inotropic effect that may be caused by both alpha- and beta-agonists. It has been suggested that 15-kDa protein phosphorylation in the heart is associated with the increase in Ca^{++} flow through slow channels of the cell membrane (Lindemann & Watanabe, 1985). However, as yet there is no convincing data concerning the spatial coupling between the voltage-dependent channels or other Ca^{++} transport systems of the cell membrane and this protein.

G Proteins

In the myocardium, calmodulin and the second messengers cAMP, cGMP, and Ca^{++} interact with G proteins that then interact directly with ion channels or with as yet unidentified regulatory components of these channels, thereby modulating transmembrane ion flow (Cockroft, 1988; Litosch, 1987; Rawis, 1987; Taylor, 1990). Stimulation of cells from chicken atria with muscarinic cholinergic receptor agonists was found to result in an increase in K^+ current into the cells and therefore hyperpolarization (Litosch, 1987). It has been established that the coupling between muscarinic receptors and K^+ channel GTP is required. Pretreatment of the cells with pertussis toxin blocked this coupling and led to ADP ribosylation of G proteins with 39-kDa and 42-kDa alpha-subunits (Litosch, 1987; Taylor,

1990). It is known that pertussis toxin prevents GTP from binding with G_i, an inhibitory G protein, to suppress the action of AC inhibitors. During this process the toxin ADP-ribosylates G_i alpha-subunits (Rawis, 1987). Infusion of $G_{pp}NH_p$ (a non-hydrolyzable GTP analogue) into frog atrial cells caused a prolonged activation of K^+ current induced by acetylcholine (Litosch, 1987). Finally, direct evidence was obtained for the functional coupling between G proteins and ion channels. Thus, a purified pertussis toxin-responsive G protein with 40-kDa alpha-subunit activated K^+ channels in the atrial cells (Litosch, 1987). These G proteins probably also directly participate in the modulation of cardiac Ca^{++} channel activity (Cockroft, 1988).

Calcium Channel Proteins in Skeletal Muscle

In adult skeletal muscle, action potentials propagate down the length of the fiber via the surface membrane and then radially into the cell interior via the transverse tubular membrane (T tubules). Each T tubule establishes specialized junctions with two terminal cisternae of the SR to form a triad. Depolarization of the T tubules results in activation of voltage-gated calcium channels in the surface membrane and influx of calcium into the muscle fiber. However, the role of voltage-gated Ca^{++} channels is not well understood in skeletal muscle. Although Ca^{++} entry into skeletal muscle from the exterior through voltage-gated Ca^{++} channels is not required for initiating excitation-contraction coupling, Ca^{++} channel antagonists have been clearly shown to block contraction in mouse myotubes (Wray & Gray, 1977). It has been shown that receptors for Ca^{++} channel antagonists are present in T tubule membranes (Catterall, 1991; Romey et al., 1989). Therefore, there is growing evidence that Ca^{++} channel antagonist receptors in the T tubule are involved in excitation-contraction coupling as voltage sensors and not as Ca^{++}-transporting channels (Brum et al., 1988; Romey et al., 1989). Physiological studies indicate that voltage-gated Ca^{++} channels in skeletal muscles are regulated by cAMP. The effects of the beta-adrenergic agonist isoproterenol on opening nitrendipine-sensitive Ca^{++} channels were additive to those of depolarization. Alprenolol, a potent beta-adrenergic antagonist, prevented the effect of isoproterenol on Ca^{++} uptake (Schimid et al., 1985). Furthermore, treatments that resulted in an increased level of cAMP in cultured skeletal muscle cells (e.g., inhibition of PDE activity with theophylline) also resulted in an increased rate of Ca^{++} entry via Ca^{++} channels (Schimid et al., 1985).

It is likely that phosphorylation by A-kinase activates a substantial number of previously inactive purified Ca^{++} channels in skeletal muscle; this suggests that those Ca^{++} channels that are active without phosphorylation in vitro were previously activated by phosphorylation in vivo and retained their incorporated phosphate through purification and reconstitution (Nunoki et al., 1989). The results focus attention on the beta-subunit of the Ca^{++} antagonist receptor as a likely site of regulation of Ca^{++} channel function by cAMP-dependent phosphorylation (Curtis & Catterall, 1985). In mammalian fast skeletal muscle, beta-agonists increase twitch tension and duration. These effects are mediated by an increase in cAMP after stimulation of beta-receptors coupled to AC localized in the T tubule membrane (Gonzales-Serratos et al., 1981). In view of the possible physiological role for phosphorylation of the Ca^{++} antagonist receptor of T tubules by A-kinase, Curtis and Catterall (1985) estimated the rate of phosphorylation of T tubule Ca^{++} channels in vivo. The results indicated that all the Ca^{++} antagonist receptors could be phosphorylated in 17 s (Curtis & Catterall, 1985). The authors concluded that the reaction should certainly occur rapidly enough to play a role in the response to beta-agonists.

The L-type channels, which are the principal voltage-gated Ca^{++} channels in muscle cells, are a complex of five protein subunits. The $alpha_1$-subunit is the central functional component of the complex and contains the receptor sites for Ca^{++} channel antagonists (Catterall, 1991). It has been proposed that the 212-kDa form of $alpha_1$ may be specialized for Ca^{++} conductance whereas the 175-kDa form may serve as the voltage sensor for excitation-contraction coupling (De Jongh et al., 1989).

The cAMP-dependent phosphorylation of purified Ca^{++} channels increases the number that are active in ion conductance (Catterall, 1991). Thus, phosphorylation-dephosphorylation can regulate the ability of the $alpha_1$-subunit to serve as functional ion channels. All six of the conserved consensus sites for A-kinase phosphorylation on the $alpha_2$-subunit are located in the 212-kDa form, and at least three of these are not present in the 175-kDa form (Catterall, 1991; De Jongh et al., 1989). Thus, a change in the relative amounts of 212 and 175 kDa may modify the regulation of the subunit by protein phosphorylation and thereby modulate Ca^{++} channel function. In addition, the beta-subunit of the L-type Ca^{++} channel is also phosphorylated to the extent of 1 or 2 mol/mol by A-kinase. A portion of this phosphate is incorporated into a site having sequence similarity to the sites in two different phosphoprotein phosphatase inhibitors (inhibitor 1 and DARPP-32) whose phosphorylation is required for inhibitory activity (De Jongh et al., 1989). Phosphorylation of this site may play a direct role in regulation of the functional activity of Ca^{++} channels, and the similarity of this site to sites on the phosphoprotein phosphatase inhibitors may allow it to function in another aspect of cellular regulation.

The physiological importance of Ca^{++} channel phosphorylation in skeletal muscle is also evidenced by experimental in situ findings. It was shown by Schimid et al. (1985) that skeletal muscle Ca^{++} channels were activated by cAMP in intact cells. It has been established that the Ca^{++} channel can be inactivated in vivo by dephosphorylation (Reuter, 1983) and that it is efficiently dephosphorylated by calcineurin, a Ca^{++}-calmodulin-dependent phosphatase (Hosey et al., 1986). Calcineurin is known to be present in relatively high levels in skeletal muscle (Stewart et al., 1982). Therefore it seems reasonable to suggest the physiological significance of Ca^{++} channel regulation by means of phosphorylation-dephosphorylation reactions.

These last results suggest that a more detailed study of the skeletal muscle SL proteins might yield important information about previously unknown functional systems in skeletal muscle membranes. In the SL of skeletal muscle, substrates for A-kinase of 15, 20, 30, 52, 54, 115, 130, 135, and 140 kDa have been found (Carafoli, 1987). It is likely that these proteins are possible membrane targets for hormones and neurotransmitters (Dolphin, 1991).

Calcium ATPase and Sodium-Calcium Carrier

Besides the cell membrane channels, there are at least two other systems for second messenger-regulated Ca^{++} transport through this membrane in the myocardium: a specific ATPase and a Na^{+}-Ca^{++} carrier.

Regulation of Calcium Transport in Cardiac Muscle

Whereas slow Ca^{++} channels in the myocardium participate in the initiation of systole by allowing trigger amounts of Ca^{++} to enter th sarcoplasm, which subsequently leads to Ca^{++} release from the SR, SL Ca^{++}-A1 ?ase—characterized by a high affinity to Ca^{++} ions but a low catalyzed reaction rate (Carafoli, 1987; Caroni & Carafoli, 1983)—seems to remove Ca^{++} from the cell during diastole.

The direction in which the third system (Na^{+}-Ca^{++} carrier) acts has already been determined. Although earlier it was believed that the carrier only removes Ca^{++} from the myocyte (Wollenberger & Will, 1978), the direction of carrier-stimulated ion flow seems to vary during the relaxation-contraction cycle in the myocardium (Carafoli, 1987; Ebashi, 1984). The carrier is characterized by a relatively low affinity for Ca^{++}, whereas the maximum carrier-stimulated Ca^{++} transport rate is approximately 30 times as high as the maximum Ca^{++}-ATPase-stimulated rate (Carafoli, 1987; Caroni & Carafoli, 1983). The carrier is electrogenous, and the stoichiometry of 3 Na^{+}/1 Ca^{++} exchange is probable (Caroni & Carafoli, 1983).

The electrogenousness of the carrier suggests that its performance is directed towards the Ca^{++} inlet into the myocyte during the plateau phase of the action potential in the heart (Barry & Smith, 1984). The rate of Na^{+}/Ca^{++} exchange seems to depend not only on concentration gradients but also on the membrane voltage. Thus, another way for Ca^{++} to flow into the myocyte is provided in addition to the calcium channels. The Na^{+}-Ca^{++} exchange allows removal from the sarcoplasm of not only those Ca^{++} ions that were brought into the cell during depolarization by the carrier itself, but also those that enter the cell through the Ca^{++} channels (Powell & Noble, 1989).

There is still little information available concerning the influence of phosphorylation and dephosphorylation on Na^{+}-Ca^{++} carrier function in the heart. For instance, cAMP-dependent phosphorylation could not be observed to produce any effect on the carrier function (Rinaldi et al., 1982). Calmodulin does not influence the carrier directly. Experiments using adenosine 5'- [gamma thio]triphosphate demonstrated that the calmodulin carrier-activating effect was associated with Ca^{++}-calmodulin-dependent protein kinase, which increases both carrier affinity to Ca^{++} and V_{max}, whereas the calmodulin-induced inactivating influence was due to the processing of SL preparations with phosphorylase phosphatase (Caroni & Carafoli, 1983). Phosphorylase phosphatase is a sarcoplasmic enzyme and is therefore unlikely to dephosphorylate SL proteins in vivo. The phosphorylase phosphatase-induced effect on carrier activity can be reproduced by incubation of SL vesicles with Mg^{++}, Ca^{++}, and calmodulin. It has been shown that membrane-bound Ca^{++}-Mg^{++}-calmodulin-dependent protein phosphatase dephosphorylated SL phosphoproteins without the obligatory involvement of the Na^{+}-Ca^{++} carrier itself (Carafoli, 1985).

The fact that Ca^{++}-ATPase is present in the cardiac SL was established in 1980 (Caroni & Carafoli, 1980). The purified enzyme has a molecular mass of 140 kDa. The SL Ca^{++}-ATPase is vanadate sensitive and has high affinity for Ca^{++} (K_m approximately 0.5 μM) and a relatively low rate of Ca^{++} transport (V_{max} approximately 0.5 nmol per mg membrane proteins at 30 °C) (Carafoli, 1985; Carafoli, 1987; Caroni & Carafoli, 1983). Unlike that performed by the Na^{+}-Ca^{++} carrier, Ca^{++}-ATPase-induced Ca^{++} removal from the sarcoplasm seems to be stimulated by cAMP-dependent processes (Carafoli, 1985; Lamers, 1985; Ziegelhoffer et al., 1979). The Na^{+}/Ca^{++} exchange probably regulates cardiac cytosolic Ca^{++} levels during excitation, whereas a specific Ca^{++}-ATPase performs this function at rest (Carafoli, 1985; Carafoli, 1987). It is likely that the influence of cAMP on the SL Ca^{++}-ATPase is mediated via the change in membrane-bound Ca^{++}. This hypothesis finds support in the data concerning the increase in Ca^{++} binding with the cardiac SL in the presence of cAMP and exogenous A-kinase (Akerman, 1982; St-Lovis & Sulakhe, 1979). However, under conditions of high protein kinase concentration, inhibition of SL Ca^{++} pump activity has been observed (Velema et al., 1983). It is likely that high phosphorylation levels in the SL are associated with a lowering of the Ca^{++}-ATPase-mediated Ca^{++} transport rate. This may lead to a prolongation of elevated Ca^{++} concentration, which may be of great importance for the development of the positive inotropic effect caused by catecholamines (Lamers & Stinis, 1980). It is not yet clear which particular mechanism underlies the phosphorylation-dependent diversity of directions in which Ca^{++}-ATPase activity is regulated: whether this mechanism is associated with the existence of several, activating and inhibiting, centers of phosphorylation in the Ca^{++}-ATPase molecule or with consecutive phosphorylation of different proteins closely connected with ATPase.

Activation of the SL Ca^{++} pump is induced not only by exogenous A-kinase but also by an endogenous, membrane-bound cAMP-dependent enzyme (Lamers & Stinis, 1980; Lamers & Weeda, 1984). A deactivating phosphatase has not been found yet. In the cardiac SL, a 9-kDa protein associated with the SL Ca^{++} pump was identified. This protein can be phosphorylated at different sites by A-kinase and calmodulin-dependent protein kinase (Lamers, 1985; Wegener & Jones, 1984). The 9-kDa protein is a monomer of a phospholamban-like 23-24-kDa protein. The 9-kDa protein as well as the 15-kDa protein takes up labelled phosphate during heart perfusion with catecholamines (Huggins & England, 1983). Lamers (1985) demonstrated that the 9-kDa protein was phosphorylated 5-10 s before an inotropic effect appeared. These findings provide evidence that 9-kDa and 15-kDa protein phosphorylation in the cardiac SL may modulate the processes of Ca^{++} transport. Phospholamban-like protein subunit phosphorylation in the cardiac SL resembles the situation in the SR, where cAMP-dependent phosphorylation and calmodulin-dependent phosphorylation produce an additive effect while stimulating the Ca^{++} pump.

It has been established that the phosphorylated 9-kDa protein is not bound with SR membranes but belongs exclusively to the outer cell membrane (Capony et al., 1983; Huggins & England, 1983; Lamers & Stinis, 1983). Under conditions of low Ca^{++} concentrations (0.3 μM), the 9-kDa protein is probably phosphorylated by A-kinase because calmodulin-dependent protein kinase displays minimum activity under these conditions (Lamers, 1985; Lamers et al., 1981). In general, it can be suggested that in situ activation of cardiac Ca^{++}-ATPase in the presence of the Ca^{++}-calmodulin complex is not associated with Ca^{++}-calmodulin-dependent SL phosphorylation (Dixon & Haynes, 1989). It is also known that the cardiac SL 9-kDa protein is phosphorylated by C-kinase (Iwasa & Hosey, 1984). The physiological importance of this phosphorylation remains unclear.

Regulation of Calcium Transport in Skeletal Muscle

Severson et al. (1972) were the first to report that a highly purified rabbit muscle SL possessed an ATP-dependent Ca^{++} accumulation ability. Mg^{++}-dependent Ca^{++}-stimulated ATPase activity is also present in skeletal muscle SL (Sulakhe & Drummond, 1974; Sulakhe & St. Lovis, 1980). It has been proposed (Sulakhe & St. Lovis, 1980) that SL Ca^{++} efflux activity only assists the SR pump in lowering the cytosolic Ca^{++} during muscle relaxation and by itself would be incapable of bringing about relaxation. The independent physiological role of skeletal muscle SL ATPase is still unclear. One possibility is that SL Ca^{++} stores participate in such key metabolic reactions in skeletal muscle as glycogenolysis. It is likely that a portion of the Ca^{++} pool, which activates phosphorylase kinase, comes from the SL Ca^{++} pool (Yeaman & Cohen, 1975). ATP-dependent Ca^{++} uptake into skeletal muscle SL fraction was shown to be increased by endogenous A-kinase phosphorylation of a 50-kDa protein (Sulakhe & Drummond, 1974). Besides this 50-kDa protein, a protein of about 30-kDa was also phosphorylated.

It is known that skeletal muscle SL passively binds Ca^{++} at two sites characterized by high and low affinities towards Ca^{++}. It is likely that phosphorylation promotes passive binding to the low-affinity site (Sulakhe & St. Lovis 1980). Thus, phosphorylation of skeletal muscle SL proteins may be of significance in altering active Ca^{++} transport and/or passive binding of this ion.

Chapter 2

cAMP-Dependent Phosphorylation of Sarcoplasmic Reticulum Proteins

Phosphorylation of intracellular membrane proteins is an important regulatory mechanism in cardiac muscle. In particular, regulation of ion flux across the sarcoplasmic reticulum is vital in the control of the contractile characteristics of the heart. Phospholamban is a key sarcoplasmic reticulum protein involved in this process. This chapter reviews these topics and discusses the phosphorylation of SR proteins in skeletal muscle.

Phosphorylation of Myocardial Phospholamban

As mentioned previously, in the SR and SL, through the participation of different second messenger systems, phosphorylation of regulatory target proteins is realized; this results in a change in ion permeability of myocyte membranes. It has already been shown that in the presence of cAMP and protein kinase, cardiac microsomes enriched with SR fraction are to a greater extent capable of ATP-dependent accumulation of Ca^{++}. Several groups of researchers, although they did not study membrane orientation of vesicles, compared Ca^{++} accumulating activity in the SR with Ca^{++} removal from the cytoplasm in vivo (Kirchberger et al., 1974; Tada et al., 1974; Wray & Gray, 1977). The influence of catecholamines on cardiac contractility results from a number of biochemical reactions that couple excitation with contraction. As a main intermediate link in this chain of reactions, cAMP-dependent phosphorylation of an SR membrane-bound protein has been distinguished.

This protein, referred to as *phospholamban* by Katz et al. (1975), forms about 4% to 5% of the total protein content in the cardiac SR. The A-kinase catalytic subunit phosphorylates purified phospholamban up to the level of 5 mol P_i per mol protein (Jakab & Kranias, 1988) (the molecular mass of phospholamban was assumed to be 27 kDa). Phospholamban is preferably phosphorylated by type II A-kinase from the heart (Tada & Inui, 1983). Introduction of the beta-agonist, isoprenaline, resulted in a fivefold (Miyakoda et al., 1988) increase in phospholamban phosphorylation by A-kinase in the heart or a sixfold increase within 60 s (Wegener et al., 1989). Beta-antagonists such as propranolol blocked this process, phospholamban being rapidly dephosphorylated (Walaas et al., 1988). In SR vesicles, cAMP-dependent phospholamban phosphorylation led to a twofold to threefold increase in Ca^{++} accumulation without changing the transport stoichiometry (2 mol Ca^{++} per mol hydrolyzed ATP) (Tada et al., 1974). Thus, in response to even insignificant adrenergic agonist concentrations, for instance 3×10^{-9} M isoprenaline, rapid (within 20 s after catecholamine introduction) phosphorylation of myocardial phospholamban occurs (Lindemann et al., 1983). During this process, phospholamban phosphorylation and SR Ca^{++}-ATPase activity undergo changes parallel with the acceleration of relaxation. Upon removal of isoprenaline from the perfusate, certain differences could be observed in the kinetics of changes between biochemical (phosphorylation) and mechanical parameters (Lindemann et al., 1983). It is likely that phospholamban phosphorylation initiates the relaxing effect produced by catecholamines, whereas the end-phase of relaxation is associated with another mechanism that differs from phospholamban phosphorylation.

Phospholamban consists of five identical subunits (Colyer & Wang, 1991; Jakab & Kranias, 1988), each of them containing 52 amino acid residues and having a molecular weight of 6,080 (Chiesi & Gasser, 1988). It is believed that phospholamban forms an ion channel consisting of five identical noncovalently bound monomers subjected to phosphorylation (Simmerman et al., 1986). A hydrophobic site at the C-end of the molecule, which is probably incorporated into the SR membrane and can interact with the hydrophobic Ca^{++}-ATPase domain, imparts typically lipophilic properties to phospholamban. In contrast, at the N-end of the molecule there is a hydrophilic domain, consisting of about 30 amino acid residues, that projects over the SR membrane. The Ser-16 and Thr-17 residues of this domain are selectively phosphorylated by A-kinase and B-kinase, respectively (Wegener et al., 1989). A serine residue is phosphorylated initially; this is followed by phosphorylation of a threonine residue (Wegener et al., 1989). In the authors' opinion, these observations suggest that (a) A-kinase, which is activated in response to beta-agonists, phosphorylates Ser-16 directly and (b) increase in intracellular Ca^{++}, resulting from adrenergic activation of the slow inward Ca^{++} channel of the SL, subsequently activates B-kinase, leading to phosphorylation of Thr-17 and a further increase in the rate of Ca^{++} uptake. Such a sequence of events, however consistent it may seem at first sight, is unlikely to reflect the situation in situ.

At the N-end of the phospholamban molecule, several positively charged amino acid residues (3 Arg and 2 Lys per subunit) are situated; these impart a basic character to the molecule. Indeed, phospholamban's pI value is about 10 (Jones et al., 1985). Phospholamban phosphorylation is accompanied by a strong acidic shift of pI value, which reaches 6.2 to 6.4 after phosphorylation of this protein by A-kinase or B-kinase. Simultaneous phospholamban phosphorylation by both kinases leads to a further acidic shift in the subunit pI values up to 5.2 (Chiesi & Gasser, 1988). Phosphorylation of Ser and Thr blocks the inhibiting phospholamban effect on Ca^{++}-ATPase due to the interaction of negatively charged phosphate groups with Lys and Arg. The latter event reduces hydrophobic interaction between phospholamban and Ca^{++}-ATPase (Kim et al., 1990).

The Ca^{++}-Mg^{++} ATPase in the SR membranes is closely connected with phospholamban, the stoichiometry of this complex approaching 1:2 (Colyer & Wang, 1991). Upon SR vesicle solubilization with either Triton X-100 or deoxycholate, phospholamban remains firmly bound to the Ca^{++} pump as demonstrated by isoelectric focusing or sucrose-gradient centrifugation (Le Peuch et al., 1980). Direct evidence was obtained for the interaction between ATPase and phospholamban in a study in which the regulatory influence of protein kinases was reproduced with the use of anti-phospholamban-specific monoclonal antibodies (Suzuki & Wang, 1986). This strong interaction between ATPase and phospholamban corresponds well with the hypothesis about the changes in the hydrophobic microenvironment of the Ca^{++} pump during phospholamban phosphorylation (Hicks et al., 1979).

Phosphorylation of phospholamban derived from the canine myocardial SR by A-kinase leads to a change in the surface membrane voltage of 7 mV. During this process, the apparent K_m value of the calcium pump increases (Chiesi & Schwaller, 1989). Such data suggest the participation of electrostatic forces in the regulation of the myocardial SR calcium pump by phospholamban. It should be noted that work on the influence of cAMP-dependent phospholamban phosphorylation on the rotating mobility of canine cardiac SR ATPase provided support for a conclusion about the association of both dephosphorylated and phosphorylated phospholamban with ATPase (Fowler et al., 1989).

Regulatory complex isolated from the myocardium after solubilization with deoxycholate consists of a polypeptide chain of Ca^{++}-ATPase, phospholamban, A-kinase, and phospholamban kinase. The latter enzyme is Ca^{++}-calmodulin dependent. Phospholamban kinase is closer to multifunctional type II B-kinase in its properties than to specific calmodulin-dependent type I protein kinases (Jett et al., 1987; Schulman & Hanson, 1993). Addition of Ca^{++} and exogenous calmodulin to cardiac microsome preparations results in phospholamban phosphorylation to the same level as achieved with incubation with catalytic A-kinase subunit (Cavadore et al., 1981). These two types of phosphorylation are characterized by an additive effect; this indicates the existence of different sites phosphorylated by cAMP-dependent and Ca^{++}-calmodulin-dependent membrane-bound protein kinases. The maximum amount of phospholamban phosphorylated by type II B-kinase under conditions of optimal Ca^{++} concentration (5-10 μM) is equal approximately to the amount of protein phosphorylated by A-kinase (Tada et al., 1983).

Additive and isolated action of A-kinase and B-kinase on Ca^{++} transport in the myocardial SR may be caused by different lipid and/or protein microenvironments of phospholamban in different reticulum fractions. Thus, calmodulin-dependent phospholamban phosphorylation can be observed mainly in terminal

SR cisterns, whereas in longitudinal SR tubules phospholamban is phosphorylated mostly by A-kinase (Gasser et al., 1988).

Evidently, these reticulum sections contain different phospholamban pools and are controlled by different regulation systems. Indeed, although phospholamban subunits have certain sites that are phosphorylated by either A-kinase or B-kinase, SR phosphomembrane analysis could not reveal any phospholamban molecules in the reticulum that could be simultaneously phosphorylated by the two kinases. This finding provides evidence for the uneven intracellular distribution of phospholamban kinase and A-kinase. Phospholamban kinase, a membrane-bound enzyme, phosphorylates its own phospholamban pool.

The cAMP-dependent phosphorylation of another phospholamban pool is to a great extent caused by a soluble form of exogenous A-kinase. Phospholamban domains phosphorylated by endogenous B-kinase may be inaccessible to exogenous A-kinase (Gasser et al., 1988). Therefore in the presence of both protein kinases, phospholamban phosphorylation is usually additive and different pools of this protein are phosphorylated. When A-kinase is present at very high concentrations together with calmodulin, the level of phosphorylation becomes more than additive; this indicates that new phosphorylation sites are appearing (Chiesi & Gasser, 1988). It is likely that the phospholamban pool interacting with phospholamban kinase, upon calmodulin-dependent phosphorylation, can undergo additional phosphorylation by exogenous A-kinase. In this case, subunits of pI 6.4 disappear and a protein emerges that is phosphorylated by the two protein kinases with pI 5.2 (Chiesi & Gasser, 1988). These findings may provide an explanation for the additional stimulating effect of A-kinase on Ca^{++} transport in the SR, an effect that can be observed in vitro (Le Peuch et al., 1980). Phospholamban phosphorylation in the presence of calmodulin and cAMP causes in particular an increase in resistance to proteolysis in SR Ca^{++} transport (Antipenko et al., 1989). It should be noted that synergistic A- and B-kinase effects in the SR have been shown only under conditions of high A-kinase concentration (not less than 500 IU/ml) (Chiesi & Gasser, 1988); this seems to rule out the physiological importance of this effect (Colyer & Wang, 1991). On the other hand, cardiac SR Ca^{++} pump activity can be regulated by different kinases owing to the compartmentalization of phospholamban (Gasser et al., 1988).

It has been established that phospholamban contains a considerable number of phosphorylation sites (up to 10) (Wegener & Jones, 1984). A wide variety of phospholamban isoforms in the cardiac muscle cell also make this protein a convenient substrate for different protein kinases. In the guinea pig heart, 11 phospholamban forms could be found; these differed from one another in their electrophoretic mobility and corresponded with protein pentamers in which from 0 to 10 sites were phosphorylated (Wegener et al., 1989). Phosphorylation of phospholamban by cGMP-dependent processes has been found to stimulate Ca^{++} accumulation in SR vesicles in vitro (Raeymaekers et al., 1988). Hawthorne and Simmonds (1989) demonstrated cardiac phospholamban phosphorylation by G-kinase, performed at the same Ser sites as cAMP-dependent phosphorylation.

Interestingly, phospholamban appeared to be an even better substrate for G-kinase than for A-kinase (Huggins et al., 1989). At the same time, under the influence of compounds causing an increase in intracellular cGMP level (guinea pig heart perfusion with carbamylcholine or 8-Br-cGMP), no increase in phospholamban phosphorylation in situ could be demonstrated (Huggins et al., 1989). One explanation for this phenomenon may be that heart A-kinase content is about 10 times as high as that of G-kinase (Murray et al., 1989). In the heart, cGMP-dependent phospholamban phosphorylation can be hindered also by the presence in the close vicinity of phospholamban of a specific phosphatase or by localization of PDE, which degrades cGMP and blocks G-kinase activation, in the same compartment.

It was also found that phospholamban could be phosphorylated by Ca^{++}-activated phospholipid-dependent protein kinase at a site that differed from the sites of phosphorylation by cAMP-dependent and Ca^{++}-calmodulin-dependent protein kinases (Iwasa & Hosey, 1984). Phospholamban phosphorylation by C-kinase, in the authors' opinion, might be activated by beta-adrenergic agonists that caused Ca^{++} entry into the cell. At the same time, alpha-adrenergic stimulation, which could result in C-kinase activation caused by the increase in diacylglycerol level, was observed not to lead to phospholamban phosphorylation in the intact heart (Lindemann, 1985). There is no unanimity as yet about the physiological role of the Ca^{++}-calmodulin-dependent phospholamban phosphorylation system.

It is believed that a permanent Ca^{++} sensor that activates the SR Ca^{++} pump as cytosolic Ca^{++} contents reach micromolar concentrations during systole is identifiable with Ca^{++}-calmodulin-dependent phospholamban kinase (Cavadore et al., 1981). At first glance, a permanent Ca^{++} pump activity in the myocardium may be caused by heart muscle function itself (it is in the myocardium that Ca^{++}-calmodulin-dependent regulation via phosphorylation is present, as is not the case with fast skeletal muscles). It has been suggested that a group of phospholamban molecules can remain phosphorylated even in the absence of beta-adrenergic stimulation, when cAMP concentration is fixed at a low basal level. In this case, Ca^{++} entry into the SR and cardiac rhythm may be controlled by calmodulin-dependent phospholamban phosphorylation (Cavadore et al., 1981). However, the coupling of contraction cycles with phospholamban phosphorylation and dephosphorylation according to kinetic parameters is hardly possible in the heart of homoiothermal animals (Lindemann et al., 1983). It should be noted that this coupling has been established for regulatory proteins of the myofibrils. At the very low pulsation rate of the tortoise heart (4 beats/min at 9 °C), the degree of myosin light chain (19 kDa) phosphorylation correlates with the development of tension (Barany et al., 1981).

Karczewski et al. (1987) observed a simultaneous increase in cAMP-dependent and calmodulin-dependent phospholamban phosphorylation in response to isoprenaline injection. The authors of this article suggest that in this case cAMP-dependent phosphorylation of regulatory membrane proteins led to the increase in cytosolic Ca^{++} levels and phospholamban kinase stimulation. Thus, calmodulin-dependent phospholamban phosphorylation can only enhance the cAMP-dependent response to beta-agonists and does not play any independent role in the heart (Karczewski et al., 1987). The "independence," but also the regulatory role of Ca^{++}-calmodulin-dependent phospholamban phosphorylation in vivo in general, is in doubt. According to Colyer and Wang (1991), incorporation of the second P_i residue into the phospholamban molecule under A- and B-kinase-induced phosphorylation does not lead to additional activation of the Ca^{++} pump. In the intact guinea pig heart (Lindemann, 1985) as well as in the rat heart (Barany et al., 1981), compounds such as ouabain and Ca^{++} ionophores, which increase intracellular Ca^{++} concentration and induce a positive inotropic effect (but, unlike beta-agonists, did not increase cAMP level), could not phosphorylate phospholamban.

Ca^{++}-calmodulin-dependent phospholamban phosphorylation can be reproduced in vitro, and SR membranes contain phospholamban kinase. At the same time, phosphate turnover rate is low at the site phosphorylated by this kinase, and in vivo, this site remains phosphorylated even in the absence of those compounds that increase contractility. Therefore, a statement by Wegener et al. (1989) about beta-agonist-stimulated phospholamban phosphorylation by Ca^{++}-calmodulin-dependent protein kinase seems to be problematic. It is likely that phospholamban phosphorylation and myocardial relaxation can change under the influence of changes in intracellular Ca^{++} concentration and with the participation of calmodulin only when intracellular cAMP concentration is high (Mundina et al., 1989).

In myocardial SR preparations, endogenous phosphoprotein phosphatase activity was found. This enzyme participated in phospholamban dephosphorylation (Kranias, 1984; Kranias & Di Salvo, 1986). The dephosphorylation rate did not depend on the way phospholamban had been phosphorylated by either cAMP-dependent or Ca^{++}-calmodulin-dependent protein kinase (Kranias, 1984). Phospholamban phosphatase activity does not depend on Ca^{++} or calmodulin. This protein phosphatase is closer in its properties to phosphatase 1 (McDongall et al., 1991), not to phosphatase 2A, as was suggested earlier (Kranias & Di Salvo, 1986).

An important role performed by protein phosphatase 1 in the heart has become evident only recently; it was underestimated earlier because the enzyme rapidly lost its activity in hearts left in situ after death (McDongall et al., 1991). Phosphorylation phosphatase activity found in the SR supplements the regulatory cycle of Ca^{++} pump function in the heart, which is caused by phospholamban phosphorylation and dephosphorylation. Thus, it has been established that upon phosphorylation by A-kinase, inhibitor 1 completely suppresses phosphoprotein phosphatase 1 activity (Cohen, 1992). In the heart, inhibitor 1 is present in the cytosol, and its phosphorylation is stimulated by isoprenaline and inhibited by beta-antagonists. Phosphorylated inhibitor 1 efficiently suppresses phospholamban dephosphorylation in the SR (Iyer et al., 1988). These data suggest the following sequence of events under conditions of beta-adrenergic hormone action on the heart: A-kinase activation → inhibitor 1 phosphorylation → protein phosphatase 1 inhibition → slowing down of phospholamban dephosphorylation → increase in Ca^{++} uptake rate in the SR.

It has become evident that A-kinase performs additional (as regards the direct phospholamban phosphorylation) regulation of Ca^{++} transport in the cardiac SR, not only by modulating the activity of inhibitor 1 but also via the change in phospholamban phosphatase compartmentalization (McDongall et al., 1991). Thus, it has been established that protein phosphatase 1 associated with the SR in the heart and skeletal muscle closely resembles or is even identical to protein phosphatase 1_G, an enzyme associated with glycogen particles (Cohen, 1992). Phosphatase 1_G, probably a main phospholamban phosphatase of the heart, consists of a catalytic subunit (about 37 kDa) and a G subunit (about 160 kDa), the latter being responsible for binding to the SR and glycogen particles (McDongall et al., 1991). The G subunit is phosphorylated by A-kinase. Phosphorylation triggers phosphatase 1_G dissociation and free catalytic subunit translocation from the SR or glycogen particles into the cytosol. The released catalytic subunit is 5 to 10 times less active than the nondissociated enzyme. Hence, in the heart, cAMP-dependent G subunit phosphorylation appears to be an important mechanism of phosphatase 1 inhibition and the corresponding additional increase in phosphorylation of Ser-16 on phospholamban in response to beta-agonists (Cohen, 1992).

Muscarinic receptor agonists, in contrast, in some way as yet unknown activate phospholamban dephosphorylation. Thus, acetylcholine significantly suppresses the phosphate uptake into phospholamban induced by isoprenaline and simultaneously blocks a positive inotropic effect produced by this catecholamine (Watanabe et al., 1981). Such antagonistic action of muscarinic receptor agonists with respect to catecholamines is most likely not associated with the decrease in cAMP level or A-kinase activity inhibition (Watanabe et al., 1981) and seems not to depend on cGMP level (Hartzell & Titus, 1982). It is very likely that muscarinic agonist action in the cardiac SR is realized through a cyclic nucleotide-independent mechanism.

Increases in Ca^{++} uptake rate in the cardiac SR in the presence of cAMP explain two main effects exerted by catecholamines on myocardial mechanical activity: systolic shortening and increased contractility. Under conditions of cAMP-dependent phospholamban phosphorylation, the acceleration of Ca^{++} uptake in the SR may cause the shortening of systole because Ca^{++} will be removed from troponin C at an increased rate. Acceleration of calcium uptake leads to an increase in Ca^{++} accumulation in the SR, thus allowing the cell to keep a certain additional amount of Ca^{++}. This additional Ca^{++} may be delivered to myofibrils during systole, thus causing the increase in myocardial contractility.

The influence of cAMP on ATP-dependent Ca^{++} transport in the myocardial SR is generally accepted, perhaps with the single exception of England et al. (1984), who did not observe phospholamban phosphorylation during heart perfusion with catecholamines. The data on the participation of calmodulin in the regulation of Ca^{++} pump activity in the cardiac SR, however, are quite contradictory. On the one hand, phospholamban is reported to bind the Ca^{++}-calmodulin complex, hindering phospholamban phosphorylation (Molla et al., 1985). This mechanism may be acting during systole, promoting the contraction under conditions of increases in intracellular Ca^{++} and blocking the stimulation of the SR Ca^{++} pump. According to this concept, cAMP-dependent phospholamban phosphorylation is realized only when the intracellular Ca^{++} value is low, that is during diastole (Lindemann, 1985; Molla et al., 1985). On the other hand, low calmodulin content in the SR membranes may rule out calmodulin's regulatory role in Ca^{++} transport via the cardiac sarcoplasmic reticulum in vivo. Experimental studies of canine cardiac SR membranes treated with [^{125}I]-calmodulin revealed not only a 120-kDa complex of calmodulin with Ca^{++}-ATPase but also calmodulin complexes with phospholamban (40 kDa) and its subunits (26 kDa and 28 kDa), respectively (Charles & Bruce, 1982). At the same time, calmodulin content in the membranes isolated in the presence of Ca^{++} was only 1 mol calmodulin per 250 mol ATPase (Hartweg & Bafer, 1983). In in vitro experiments simulating the physiological conditions, the addition of saturating calmodulin concentrations to SR membrane preparations resulted in only insignificant (1.2-1.4-fold) increases in oxalate-dependent Ca^{++} accumulation (Wuytack et al., 1980). In addition, the participation of calmodulin in the regulation of myocardial SR Ca^{++} pump activity in vivo can be questioned in view of the fact that the affinity of this enzyme to calmodulin is about 10 to 100 times lower than that of most calmodulin-dependent enzymes (Orlov, 1987).

Purified phospholamban isolated from the canine heart contains about 0.6 μmol of lipid P_i per mg protein (Jakab & Kranias, 1988). The main phospholipids associated with phospholamban are phosphatidylserine (34%), phosphatidylcholine (22%), sphingomyelin (17%), phosphatidylethanolamine (9%), and phosphoinositide (PI) (13%), which stimulates phosphorylation of purified phospholamban (Suzuki & Wang, 1987). These phospholipids form a complex with phospholamban that is phosphorylated by catalytic A-kinase subunit, up to 4 nmol of phosphate being incorporated per mg protein (Jakab & Kranias, 1988). The main phosphorylated phospholipids are the phosphoinositides PIP and PIP_2. PI phosphorylation by A-kinase has also been demonstrated in rabbit cardiac SR (Enyedi et al., 1984). It is thus far unclear whether A-kinase itself directly phosphorylates phosphoinositides or whether A-kinase regulates the activity of endogenous phosphatidylinositide kinases. At any rate, catalytic subunits of A-kinase phosphorylate purified PI to form only PIP_2 (Jakab & Kranias, 1988). At the same time, phospholamban phosphorylation is associated with the formation not only of PIP_2 but also of PIP. Polyphosphoinositide phosphorylation is reversed by phospholamban phosphatase and inhibited by thermostable A-kinase inhibitor (Jakab & Kranias, 1988).

Phospholipid phosphorylation, like protein phosphorylation, can also take part in the regulation of membrane function. For instance, it is believed that PI phosphorylation regulates membrane permeability for ions (Berridge, 1984) as well as SR Ca^{++}-ATPase activity in the skeletal muscles (Varsanyi et al., 1983). It has been demonstrated that SR Ca^{++}-ATPase activity undergoes a considerable increase during phosphorylation of enzyme-associated PI to PIP and plasmalemmal Ca^{++}-ATPase activity in red blood cells (Redman, 1972). Although polyphosphoinositides are minor components as compared with PI, their phosphorylation and dephosphorylation seem to be capable of changing phospholamban's polarity and thus of modifying the conformation of this protein. These changes in the conformation of phospholamban seem to play a key role in phospholamban phosphorylation by protein kinases, in the interaction between phospholamban and Ca^{++}-ATPase, or in both.

It is possible that the effect of beta-adrenergic agonists on the mammalian heart is in part caused by the increased formation of polyphosphoinositides and by corresponding changes in the properties of the SR. Thus, beta-adrenergic stimulation of a contracting heart together with phospholamban phosphorylation at the peak of the inotropic response (Kranias, 1985) also leads to the increased phosphate incorporation into phosphoinositides, resulting in the formation of PIP and PIP_2 (Jakab et al., 1988). The influence of protein phosphorylation on Ca^{++} transport into the SR is sufficiently well studied; however, regulation of Ca^{++} release from the SR through reticulum channels in response to an electric stimulus and/or intracellular second messengers such as Ca^{++} or IP_3 is largely unclear. It is known that the ''light'' reticulum contains large amounts of Ca^{++}-ATPase and calsequestrin, a calcium-binding protein. In contrast, the ''heavy'' SR fraction with high specificity binds Ca^{++} channel blockers; this indicates the preferential participation of this fraction in Ca^{++} release into the cytoplasm in response to electric excitation of the SL or to nicotinic receptor agonists (Orlov, 1987). Kim et al. (1987) showed that phospholamban phosphorylation by Ca^{++}-calmodulin-dependent protein kinase did not influence passive Ca^{++} release from canine cardiac SR vesicles preloaded with the ion. Obviously, the mechanism of Ca^{++} release from the SR is not caused by calmodulin-dependent phosphorylation. It is believed that cAMP-dependent phosphorylation in the myocardial SR is associated with stimulation of Ca^{++} release (Kim et al., 1987). However, Kovacs et al. (1988) reported that dephosphorylated phospholamban could operate as a calcium channel. In the heavy SR fraction, high-molecular-weight proteins were found, their molecular weights being 305 kDa and 320 kDa respectively; both proteins could be phosphorylated by endogenous calmodulin-dependent and exogenous cAMP-dependent protein kinases (Orlov, 1987). The functional role of these proteins in Ca^{++} transport regulation remains unknown.

Phosphorylation in Slow and Fast Skeletal Muscle

It is well known that contractility in slow skeletal muscle is modulated by beta-adrenergic agonists whereas contractility in fast skeletal muscle appears to be relatively insensitive to catecholamines (Marsden & Meadows, 1970). Thus, catecholamines increase the rate of tension development, the maximum tension, and the rate of relaxation of slow-twitch skeletal muscles and heart, whereas in fast-twitch skeletal muscle only a slight change in tension is observed due to prolongation of the active state.

It has been suggested (Kirchberger & Tada, 1976) that the relaxation-promoting effects of epinephrine may be related to the presence of phospholamban in slow-twitch skeletal muscle microsomes, whereas the absence of such effects in fast-twitch skeletal muscle is due to the absence of phospholamban. Treatment of SR from fast-twitch rabbit skeletal muscle with cAMP and A-kinase had no effect on the initial rate of Ca^{++} uptake, and no radioactivity attributable to protein kinase-catalyzed phosphorylation of SR was apparent after SDS-PAGE (Kirchberger & Tada, 1976). At the same time a phosphate acceptor protein of 22 kDa was phosphorylated by A-kinase with increases in Ca^{++} transport activity in skeletal muscle SR of the slow-twitch rabbit soleus muscle (Kirchberger & Tada, 1976). The cAMP-produced tension relaxation developed by skinned fibers from cat caudofemoralis was not accompanied by cAMP-dependent modification of myofilament sensitivity to Ca^{++} (Fabiato & Fabiato, 1978).

Phospholamban content differs in different muscle types. For example, the phosphorylated phospholamban content in extensor carpi radialis (predominantly fast) is 12.3 pmol ^{32}P per mg protein; in superficial digitalis flexor muscle (predominantly slow), 78.3; and in cardiac muscle, 492 pmol ^{32}P per mg protein (Jorgensen & Jones, 1986). Like cardiac phospholamban, this protein appears to be a phosphorylatable oligomer that upon boiling in SDS is converted to a 5-6-kDa form (Jorgensen & Jones, 1986).

Although no phospholamban was found in fast skeletal muscle preparations, increased Ca^{++} accumulation was observed coincident with phosphorylation of a 95-kDa protein in fast skeletal muscle SR by phosphorylase kinase or A-kinase (Schwartz et al., 1976; Van Winkle & Entman, 1979). It appears that Ca^{++} transport in fast skeletal muscle SR preparations is modulated by phosphorylation of various membrane components, including the Ca^{++}-ATPase (Bornet et al., 1977; Galani-Kranias et al., 1980). In most instances, the relationship between the phosphorylation of various membrane components and their influence on Ca^{++} transport is still being debated. Fabiato and Fabiato (1978) suggested that cAMP induced an increase in the relaxation rate of skinned fast skeletal muscle only under conditions in which SR Ca^{++} uptake is diminished. This may mean that for the manifestation of the physiological effect of catecholamines via protein phosphorylation activation in the SR, conditions are required that differ from the normal Ca^{++} status of fast-twitch skeletal muscle.

In general, phospholamban seems to be a weak regulator of the Ca^{++} pump in slow skeletal muscle, even though the ratio of phospholamban to the Ca^{++} pump in this tissue is similar to that in cardiac muscle (Briggs et al., 1992).

Chapter 3

Myofibrillar Protein Phosphorylation

The phosphorylation and dephosphorylation of troponin are critical in controlling contractile processes in cardiac muscle. This chapter reviews phosphorylation of myofibrillar proteins, with special emphasis on the phosphorylation of troponin.

Troponin

Myocardial contraction results from Ca^{++} binding to troponin C, whereas the regulation of myofibrillar complex activity through second messengers is realized mainly with the participation of the inhibitory component of troponin complex (troponin I).

In cardiac muscle the largest amount of endogenous phosphate is associated with troponin I. In contrast, in skeletal muscle troponin T binds the largest fraction. The physiological importance of troponin T phosphorylation in the skeletal muscle has not been established. Cardiac muscle troponin I is phosphorylated 30 times as rapidly as that of the skeletal muscle. It should be noted that phosphorylation has been performed through use of cAMP-dependent protein kinase showing high affinity to cardiac muscle troponin I endogenous kinase, as well as to specific protein phosphatase, which appeared to be closely associated with cardiac muscle troponin (Reddy, 1976). Troponin I is the only component of cardiac troponin that can be phosphorylated by this endogenous protein kinase; the effectiveness of the phosphorylation depends on the pH. The maximum phosphorylation rate was found to occur at pH ranging from 8.5 to 9.0, while at pH ranging between 6.0 and 7.0, V_{max} was 13% and 45% of that at pH 8.5, respectively (Stull & Buss, 1977). The influence of pH on the rate of in vivo troponin I phosphorylation may have a certain functional importance. Intracellular pH level in the normal myocardium is about 7.0; the phosphorylation rate is only 45% of maximum—that is, a kind of phosphorylation ''reserve'' must exist. A slight change in pH value (as, for example, in acidosis) may lead to the change in the troponin I phosphorylation rate.

The enzyme A-kinase catalyzes phosphorylation of Ser-20 and Ser-146 residues in myocardial troponin I molecules (Moir et al., 1977). It is believed that a higher rate of cAMP-dependent cardiac troponin I phosphorylation as compared to skeletal muscle troponin I phosphorylation is caused primarily by the high rate of serine residue phosphorylation in position 20. It is this residue that is phosphorylated in vivo; in vitro its rate of cAMP-dependent phosphorylation is two to three times higher than that of nonfractional histone phosphorylation (Laurence et al., 1979). Data are available indicating that A-kinase can "recognize'' possible sites for phosphorylation in substrate molecules by the extent to which serine is surrounded by basic amino acid residues. Thus, the amino acid sequence around Ser-20 in myocardial troponin I contains the largest number (four) of arginine residues, contributing significantly to phosphorylation kinetics. Another phosphorylizable residue in cardiac troponin I, Ser-146, is situated in the close vicinity of the actin-binding site in the troponin I molecule (Kemp, 1979). This suggests a change in the protein-protein interaction rate by means of phosphorylation.

The influence of beta-agonist on the acceleration of myocardial relaxation or on the increase in myocardial contractility is realized via cAMP-dependent phosphorylation of both membrane proteins and myofibrillar proteins of cardiac cells. In the latter case, a regulatory protein of the troponin-tropomyosin complex, troponin I, is phosphorylated. Cardiac troponin I, apart from A-kinase, can also be phosphorylated by phosphorylase kinase, cAMP-independent protein kinase (Reddy, 1976), cGMP-dependent protein kinase (Lincoln & Corbin, 1978), and Ca^{++}-phospholipid-dependent protein kinase (Iwasa & Hosey, 1984). However, the most important of these processes for in vivo conditions seems to be cAMP-dependent troponin I phosphorylation. The activity of this process, unlike that of phosphorylase kinase, is not inhibited by troponin C (Adelstein, 1983).

The fact that Ca^{++}-regulated ATPase activity of myofibrils is modulated by cardiac troponin I phosphorylation is evidenced by a variety of data. Adrenaline-induced or isoproterenol-induced increases in myocardial contractility are accompanied by the growth of cAMP content, A-kinase activation, and increases in the rate of myocardial troponin I phosphorylation (England, 1975). It has been established that calcium-induced ATPase activation (up to half of maximum activity) of reconstructed actomyosin—consisting of ^{32}P-phosphorylated cardiac actin-tropomyosin-troponin complexes and cardiac myosin—occurs at higher Ca^{++} concentrations than is the case when actomyosin contains a non-phosphorylated actin complex.

Catecholamine-induced myocardial troponin I phosphorylation influences troponin subunit interaction. It is known that phosphorylated sites of troponin I and troponin T molecules are situated sterically close to each other in the sites of protein-protein interaction and also that their phosphorylation may affect this interaction and the functioning of the troponin complex as a whole (Moir et al., 1977). For instance, cAMP-dependent cardiac troponin I phosphorylation may cause a change in the Ca^{++}-binding properties of troponin C, thus modifying the responsiveness of actomyosin ATPase to Ca^{++} (Holroyde et al., 1979). It has been established that interaction with troponin C inhibits phosphorylation of troponin I sites significantly distant from each other in the primary structure (Syska et al., 1979).

Myocardial cAMP-dependent troponin I phosphorylation is obviously related to cardiac contractility. Under the influence of isoproterenol, glucagon, and dibutyryl cAMP, a rise in A-kinase activity was found in rat cardiac muscle; in this case the increase in the maximum myocardial relaxation rate exceeded that of the maximum myocardial contraction rate, which was induced by the same agents. As catecholamines ceased entering the blood flow, troponin I remained in a phosphorylated state, while the contraction amplitude returned to the initial value (England, 1975). Compounds that cause positive inotropic effects but do not increase tissue cAMP content (ouabain, ionophore X537A, etc.) do not lead to troponin I phosphorylation. It should be noted that troponin phosphorylation may also be a signal for cardiac myofibrillar protein synthesis or degradation (Gusev & Dobrovolsky, 1976).

Phosphorylation of cAMP-dependent troponin I seems to regulate the range of Ca^{++} concentrations necessary for contractile protein activation. Such phosphorylation probably leads to the reduction of cardiac myofibril responsiveness to Ca^{++} during the action of beta-agonists (Perry, 1979; Westwood & Perry, 1981). It has been suggested that in vivo the beta-adrenergic system will activate and the cholinergic system will inhibit troponin I phosphorylation (Winegrad et al., 1983). It has also been reported that an increase in cGMP level causes cardiac troponin I dephosphorylation (England et al., 1984); however, the underlying mechanism is not yet clear.

As previously mentioned, in a complete troponin complex, A-kinase phosphorylates Ser-20 residues in the troponin I molecule (both in vivo and in vitro); troponin's Ca^{++}-specific center affinity to Ca^{++} becomes reduced. This may allow troponin to exchange "regulatory" Ca^{++} more rapidly in order to work synergistically with the SR activity increasing with phosphorylation. One of the physiological functions of cAMP-dependent troponin I phosphorylation may consist of participation in this negative feedback mechanism during contraction. Thus the effect of fading away caused by the inhibition of myofibrillar Ca^{++}-ATPase activity (and consequently of the actin-myosin interaction) by troponin I phosphorylation may mitigate the hyperresponse of cardiac contractility growth under conditions of increased Ca^{++} intracellular concentration induced by catecholamines (Adelstein, 1983; Resink, 1979). It should be noted that studies of the influence of beta-adrenergic troponin I phosphorylation on the functioning of actomyosin complex have yielded contradictory results. For example, in the guinea pig heart, actomyosin ATPase responsiveness to Ca^{++} was shown to increase with troponin I phosphorylation (Rubio et al., 1975).

Experiments in vitro have established that cardiac troponin I (1.1 mol phosphate per mol of protein) and troponin T (0.6 mol phosphate per mol protein) were phosphorylated by protein kinase C (Vorotnicov et al., 1988); however, the participation of this protein kinase in troponin phosphorylation in vivo is far from obvious.

Other Substrates

Tropomyosin is phosphorylated by a specific Ca^{++}-activated tropomyosin kinase (Barany & Barany, 1980). The physiological significance of such phosphorylation is not yet clear. Phosphorylation of the thick filament's protein, a so-called C protein, has been better studied. This protein is localized with a periodicity of 43 nm along the thick filament and has a molecular mass of 150 kDa in the heart (Schlender et al., 1987). This C protein is an excellent substrate for A-kinase. During phosphorylation, phosphate uptake by this protein ranges from 3 mol (chicken heart) to 6 mol (bovine heart) of phosphate per mol of protein (England et al., 1984). The rate of C protein phosphorylation in the heart is higher than that of troponin I and lysine-rich histone. In the heart, C protein is phosphorylated in response to beta-adrenergic stimulation and dephosphorylated in response to cholinergic agonists (England et al., 1984). A correlation has been established between the extent of C protein phosphorylation and arterial blood pressure, as well as between the former and myocardial relaxation rate (England & Krause, 1987). Although in vitro, C protein is phosphorylated by B- and C-kinases, it is evident that the physiological significance of C protein phosphorylation is associated with A-kinase and manifests itself in the modulation of actomyosin Ca^{++}/Mg^{++}-ATPase activity (England et al., 1984). In vivo, C protein is most likely dephosphorylated by protein phosphatase 2A (Schlender et al., 1987). Troponin I and C protein from the rat heart, under the influence of catecholamines, were observed to remain in a phosphorylated state for several minutes (England & Krause, 1987). It is suggested that the relatively low rate of myocardial contractile protein dephosphorylation is caused by the presence in cardiac muscle cell cytosol of various phosphoprotein phosphatase inhibitors.

In the sarcoplasm of cardiac cells, a myosin light-chain kinase of 77 kDa has been found. This enzyme is highly specific in the reaction of gamma phosphate transfer from ATP to the only serine residue at the N-terminus of myosin light chain, the activity of myosin light-chain kinase (as well as that of cAMP-dependent protein kinase) being independent of Ca^{++} concentration in the medium (Chacko et al., 1977). The responsiveness of light-chain kinase to Ca^{++} is caused by calmodulin; a site for calmodulin binding has been reported in the molecule of myocardial myosin light-chain kinase (Walsh et al., 1980).

It has been demonstrated that myosin light-chain phosphorylation in perfused rabbit heart correlates with the physiological condition of the myocardium (Barany & Barany, 1980; Chacko et al., 1977). Data on the influence exerted by myosin light-chain phosphorylation on myocardial myofibril ATPase activity are rather contradictory. Myosin dephosphorylation caused a decrease in actin-activated ATPase activity (Barron et al., 1980). The addition of tropomyosin complexes from skeletal muscle significantly increased ATPase activity of phosphorylated (but not of dephosphorylated) myosin in the presence of Ca^{++} ions.

On the other hand, phosphorylation by myosin light-chain calmodulin-dependent kinase of cardiac myofibril preparations led to a 50% inhibition of ATPase activity (Franks et al., 1984). Neither catecholamines nor the increase in intracellular Ca^{++} concentration caused acceleration of phosphate inclusion into myocardial myosin light chains (phosphate half-life in light chains is about 3 min) (England, 1988). These data support the opinion that light-chain phosphorylation does not contribute to the short-term effects produced by beta-adrenergic stimulators of myocardial contraction (Barany & Barany, 1980).

In myocardial preparations a myosin V_3 isoform was found whose activity could be modulated by beta-adrenergic agonists (Winegrad et al., 1983). Thus, cAMP-dependent contraction regulation associated with myosin may exist in the heart. It is still unclear what influence catecholamines exert on the level of myosin phosphorylation in the intact myocardium and the biochemical role of light-chain phosphorylation in the functioning of cardiac muscle.

In conclusion, it should be noted that two major substrates for A-kinase in the heart are troponin I and phospholamban. Phosphorylation of phospholamban by A-kinase at Ser-16 relieves Ca^{++}-ATPase activity inhibition, thus increasing the rate at which the Ca^{++} released from troponin C is transported into SR. The combined phosphorylation of troponin I and phospholamban therefore allows beta-agonists (acting via A-kinase) to lower the cytosolic concentration of Ca^{++} more rapidly, thereby increasing the rate of relaxation of cardiac muscle. This mechanism operates only in cardiac muscle, because (a) phospholamban is not present in fast skeletal muscle and (b) skeletal muscle contains a distinct isoform of troponin I that lacks the amino-terminal region of cardiac troponin I, which contains the Ser residues phosphorylated by A-kinase (Cohen, 1992).

PART II

The Effects of Exercise on cAMP Metabolism and Properties of cAMP-Dependent Protein Kinase of Cardiac and Skeletal Muscle

Chapter 4

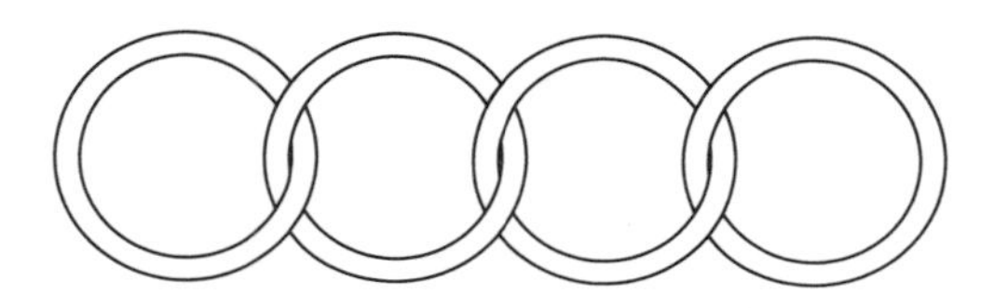

Exercise, cAMP, and cAMP-Dependent Protein Kinase in Cardiac Muscle

This chapter summarizes the available data on exercise and the cAMP system. Included is information from over 20 years of M.I. Kalinski's experimental work at the Laboratory of Muscle Biochemistry at the Institute of Biochemistry, Ukraine's Academy of Science, and the Department of Biochemistry at the Institute of Physical Culture in Kiev, Ukraine. Much of this work is virtually unknown to Western scientists, having been previously published mostly in the Slavic languages.

cAMP Metabolism

Exercise produces changes in myocardial catecholamine concentration and turnover (Mazzeo, 1991) and in beta-receptor sensitivity to catecholamines (Hammond, 1988; Takeda, 1985). These can induce intracellular changes via the second messenger cAMP in cardiac muscle during exercise. cAMP and cAMP-dependent protein kinase promote glycolysis (Namm, 1968) and lipolysis (Palmer et al., 1986, 1990) and accelerate protein synthesis (Xenophontos at al., 1989) in heart muscle. Also, cAMP increases Ca^{++} influx across the SL (Katz, 1990) and stimulates Ca^{++} transport by the sarcoplasmic reticulum (Tada et al., 1983). All of these processes are critical in the regulation of contraction of the heart. This is particularly true during exercise when the demands on the heart are intensified (Antipenko et al., 1992; Kalinski et al., 1989; Tibbets et al., 1989).

Several studies have dealt with the effects of relatively short bouts of exercise, that is, 5 to 60 min of running, on cardiac cAMP metabolism (Goldfarb et al., 1986; Palmer, 1988; Palmer & Doukas, 1983; Palmer et al., 1981), but little attention has been given to the effects of acute, prolonged exercise to exhaustion on the cAMP content of the heart (Goldfarb and Kendrick, 1981). In addition, little information is available concerning [cAMP] and cAMP-PDE activity in cardiac muscle during the postexercise period.

There are inconsistencies in the research on effects of exercise training on cAMP metabolism in cardiac muscle. Noakes et al. (1983) did not find any change in the [cAMP] of cardiac muscle of exercise-trained animals, whereas Palmer et al. (1980) reported significant increases in [cAMP] in the hearts of trained animals. Dohm et al. (1976) observed decreases in AC activity in the hearts of trained animals, but Moore et al. (1982) and Wyatt et al. (1978) did not find any changes in the basal activity of AC in the hearts of trained animals. Palmer and Doukas (1983) reported increases in cAMP-PDE activity in cardiac muscle of trained rats, but Dohm et al. (1976) did not find any significant differences.

In only one investigation were [cAMP] AC, and cAMP-PDE activities in cardiac muscle investigated both during acute exercise and at 48 hours postexercise (Kalinski et al., 1984). In this study of untrained animals, it was found that immediately after a 4 to 6 hr run, cardiac [cAMP] and AC activity were decreased 39% and 50%, respectively, as compared to the values in controls (Figure 4.1). The activity of cAMP-PDE did not change significantly. The most dramatic increases occurred during the third hr of the postexercise period, when [cAMP] and AC activity increased 61% and 38%, respectively. After this rise, [cAMP] and AC gradually declined in a nearly parallel manner until the 48th hour postexercise. The increases observed in [cAMP] and AC activity during the postexercise period at 2, 3, 6, and 12 hr were significantly greater than in the controls. In addition, [cAMP] was still significantly elevated at 24 hr. The cAMP-PDE activity increased slowly after exercise until between the 12th and 24th hr of recovery when it peaked, to return to control values by 36 hr.

The untrained animals showed a 26% decrease in the concentration of norepinephrine (NE) after a single bout of prolonged mild exercise (Figure 4.2).

The effects of exercise training on [cAMP], AC, and cAMP-PDE activities in the cardiac muscle of trained animals can be seen in Table 4.1 (Kalinski et al., 1980). Training did not significantly alter [cAMP] or cAMP-PDE activity before exercise. However, AC activity was increased, before exercise, as a result of endurance training. When the trained rats were subjected to 6 hr of running, they showed significant increases of 20% in AC activity and 35% in cAMP-PDE activity (Table 4.1). Also, the trained rats showed a 55% gain in the [cAMP] in cardiac tissue after exercise when compared to the untrained animals in the preexercise state (Table 4.1).

Exercise presents a challenge to the heart by imposing additional work, and the normal response is an augmented cardiac output. Both chronotropic and inotropic responses contribute to the increased cardiac demand via catecholamines and their receptors on the myocardium. Both beta 1- and beta 2-adrenoreceptors are coupled to the activation of AC (Nawrath, 1989). The large decreases in [cAMP] and AC activity following exercise in the untrained animals (Figure 4.1) were probably related to decrements in [NE] after exercise and were responsible, in part, for the inhibition of both AC and cAMP. Kalinski et al. (1969) showed

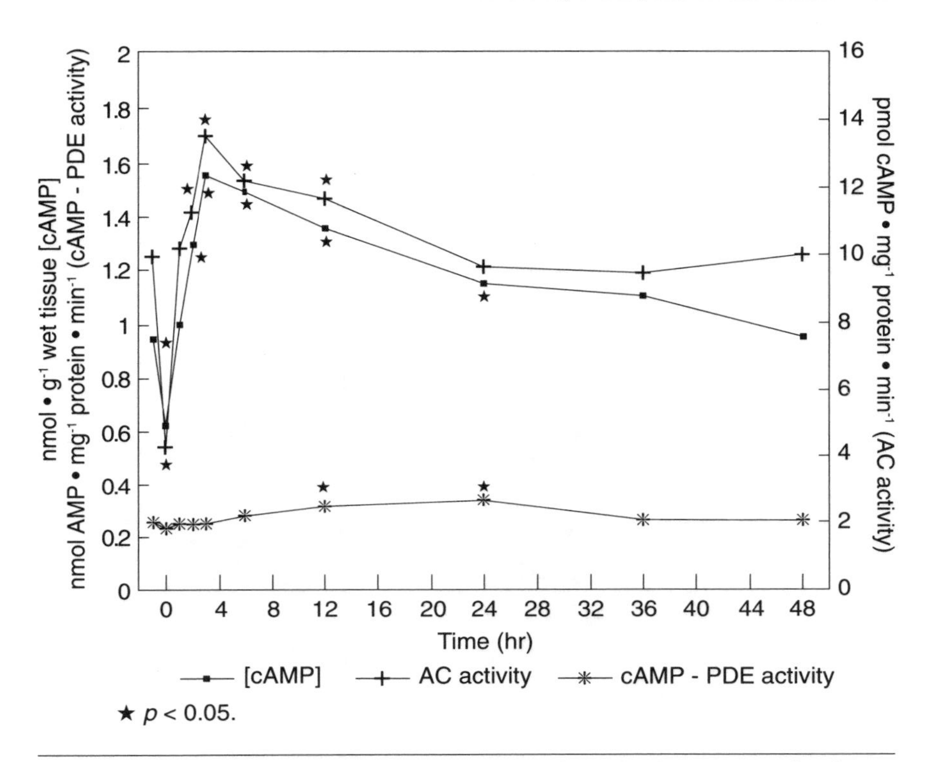

Figure 4.1 cAMP metabolism in cardiac muscle of untrained rats preexercise, immediately after exercise, and 48 hr postexercise. Data from Kalinski et al., 1984.

that acute exercise of similar intensity and duration decreased the [NE] in cardiac muscle of untrained rats (Figure 4.2). The changes in [cAMP] are in contrast to those reported by Palmer et al. (1981) and Goldfarb and Kendrick (1981) and may be related to differences in the intensities and durations of the exercise bouts. The animals in the study of Kalinski et al. (1980) were running at 14 m $\cdot$ min^{-1} for 4 to 6 hr whereas Palmer et al. (1981) reported that their rats were running at 26.8 m $\cdot$ min^{-1} for 1 hr, that is, at a 1.9-fold greater intensity but for only 20% of the exercise time used by Kalinski et al. (1984). The difference in speed obviously necessitated a change in the mode of the exercise. Goldfarb and Kendrick (1981) used a running protocol in which the speed was varied from 18.8 m $\cdot$ min^{-1} to 34.8 m $\cdot$ min^{-1} and the grade from 7.5 to 25% until exhaustion, which was at 2.7 hours. Clearly, the decreases in [cAMP] and AC activity were related to the long duration and relatively low intensity of the exercise. This observation is not true if the exercise is performed at high intensity and is sustained for less than 4 hr (Palmer et al., 1981; Goldfarb & Kendrick, 1981).

The decrease in the [NE] in the hearts of the untrained rats after exercise concurs with findings reported by Gorochow (1969), who used a 10-hr swim as the mode of exercise for his untrained rats. The exercise time remains an important

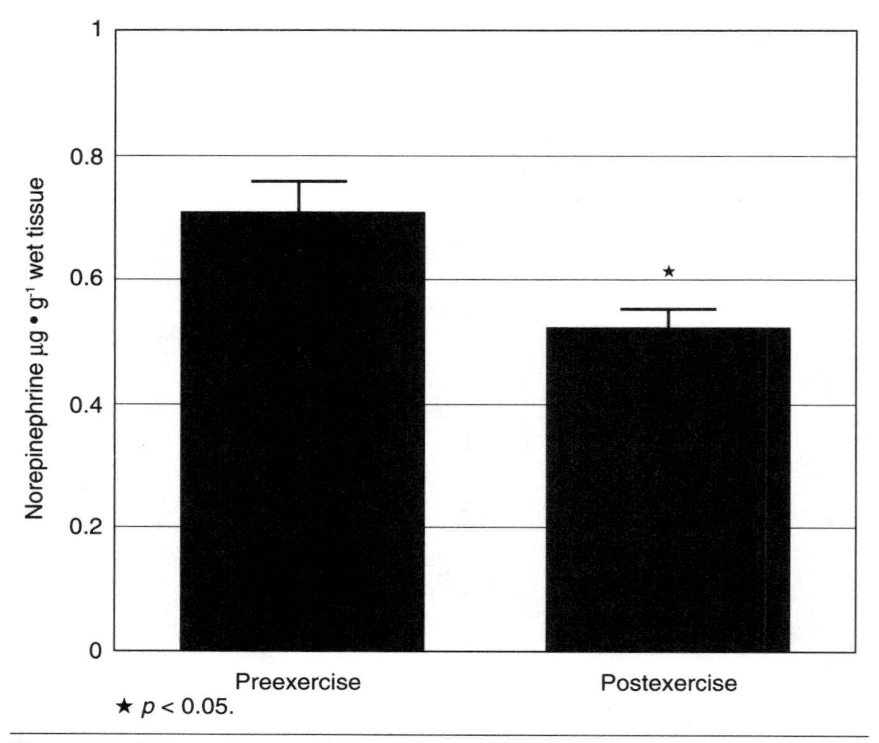

Figure 4.2 Effects of an acute bout of 4-6 hr of running on [NE] in cardiac muscle of nontrained animals, preexercise and postexercise. Data from Kalinski et al., 1969.

variable because it has been shown that exercise of 30-min duration did not change [NE] and increased NE turnover in cardiac muscle of untrained rats (Mazzeo & Grantham, 1989). A decrease in the [NE] is presumed to be caused by the decline of synthesis of NE in the presynaptic terminals of sympathetic neurons (Mazzeo, 1991). Therefore, if the exercise is of sufficient duration (i.e., 4-6 hr), [NE] can be expected to decrease immediately after exercise. It has been shown that epinephrine encapsulated in liposomes increases [cAMP] in heart, skeletal muscle, and liver (Kalinski et al., 1990). Other changes observed in this study included enhanced levels of blood glucose, free fatty acids, and ketone bodies, which resulted in an increase in the work capacity of the rats.

The lack of a significant increase in [cAMP] in the hearts of the trained animals before exercise as reported by Kalinski et al. (1980) is surprising in light of a significant elevation in AC activity (Table 4.1). Kalinski et al. (1980) also reported that trained animals showed an increase in cardiac AC activity during the preexercise period. This is not in agreement with the findings of others reporting a decrease (Dohm et al., 1976) or no change in AC activity before exercise in the trained rat (Moore et al., 1982) and cat heart (Wyatt et al., 1978). The lack of a consensus may be related to differences in training mode, species, duration, or intensity of exercise training.

Table 4.1 Effects of Endurance Exercise Training on cAMP Metabolism in Cardiac Muscle of Rats

	Untrained	Trained	
Variable	Pre-ex $n = 5$	Pre-ex $n = 5$	Post-ex $n = 5$
[cAMP] nmol · g^{-1} wet tissue	1.716 ± 0.216	2.247 ± 0.242	3.79 ± 0.216[a]
AC activity pmol cAMP · min^{-1} · mg^{-1} protein	9.99 ± 0.29	12.98 ± 0.56[b]	16.25 ± 0.47[c,d]
cAMP-PDE activity nmol AMP · min^{-1} · mg^{-1} protein	0.256 ± 0.071	0.232 ± 0.046	0.355 ± 0.012[e]

Note. Values are mean ± SE.

[a] $p < 0.01$ post-ex trained versus pre-ex untrained.

[b] $p < 0.001$ pre-exercise trained versus pre-exercise untrained.

[c] $p < 0.001$ post-ex trained versus pre-ex untrained.

[d] $p < 0.001$ post-ex trained versus pre-ex trained.

[e] $p < 0.05$ post-ex trained versus pre-ex trained and post-ex trained versus pre-ex untrained.

The data are from Kalinski et al., 1980.

In the postexercise period, [cAMP] and AC and cAMP-PDE activities have been shown to increase in the hearts of trained rats after a single exercise session (Kalinski et al., 1980) (Table 4.1). Similar findings have been reported by Palmer and Doukas (1983), who also showed that exercise increases cAMP-PDE activity in the myocardium of trained rats. These findings indicate that mild, prolonged exercise may increase the capacity of the myocardium to mobilize intracardiac energy resources by stimulation of lipolysis and glycogenolysis. The increased levels of AC and cAMP-PDE activities provide a biochemical model for cellular homeostasis. That is, the elevated AC activity increased the cellular [cAMP]; this resulted in increased glycogenolysis and lipolysis and a larger pool of energy for the working heart. In contrast, the increase in cAMP-PDE activity established a servomechanism to limit overproduction of cAMP and contribute to the economy of the cell.

Exercise training has been shown to increase beta-receptor density and binding capacity in endurance-trained skeletal muscle of rats (Buckenmeyer et al., 1990; Williams et al., 1984). There has been less agreement on beta-receptor alterations with exercise training in cardiac muscle. Some investigators have shown no changes in the beta-adrenergic system of the heart with training (Moore et al., 1982); others have found a decrease in binding capacity of beta-1 and beta-2 adrenoceptors of the sarcolemmal membranes of myocardial cells in trained rats (Werle et al., 1990). There is also evidence of an increased sensitivity of the heart of the trained animal to catecholamines at the beta-receptors and of elevated AC activity levels (Hammond et al., 1988; Wyatt et al., 1978). Since AC and beta-1 and -2 receptors are coupled (Nawrath, 1989), any alterations in the beta-receptors could influence AC activity. This could explain the up-regulation of AC activity in the trained hearts of animals both before and after exercise.

cAMP-Dependent Protein Kinase

It is known that cAMP-regulated metabolism within the cell is dependent upon activation of A-kinase. The activation of A-kinase regulates glycogenolysis (Krebs & Beavo, 1979), lipolysis (Palmer et al., 1986), protein synthesis (Xenophontos et al., 1989), Ca^{++} influx across SL (Lamers, 1985), transport of Ca^{++} into the sarcoplasmic reticulum (Katz, 1990), and phosphorylation of the myofibrillar protein troponin I. All of these biochemical pathways are dynamic during exercise and are subject to adaptations resulting from exercise training (Booth & Thomason, 1991). Endurance training increases the rate of phosphorylation of myocardial SL and SR by A-kinase in rats (Kalinski et al., 1985). These data suggest increases in both Ca^{++} flux across the myocardial cell membrane and uptake by the SR. Both adaptations could result in enhanced contractility of the endurance-trained myocardium.

As stated above, the metabolic reactions that activate Ca^{++} and cyclic nucleotide second messenger systems play a key role in the regulation of contractile properties and metabolism of cardiac muscle, including phosphorylation of proteins in myofibrils, SL, and SR (Antipenko, et al., 1992; Lamers, 1985; Opie, 1982; Tada & Katz, 1982).

The activation of A-kinase by cAMP regulates enzymes such as phosphorylase kinase, phosphorylase, and lipoprotein lipase, which when converted to their active forms stimulate glycogenolysis and lipolysis (Goldberg & Khoo, 1985; Krebs & Beavo, 1979; Palmer, et al., 1986). Also, Ca^{++} influx across the SL (Lamers, 1985; Tada & Katz, 1982) is regulated by these protein kinases as is the transport of Ca^{++} into the SR (Tada & Katz, 1982). The myofibrillar protein troponin I is controlled by A-kinase, and protein synthesis within the myocardial cell appears to be regulated by these kinases (Watson et al., 1989; Xenophontos et al., 1989). It has been shown that exercise training affects all of these biochemical pathways in cardiac muscle (Borensztajn et al., 1975; Goldfarb & Kendrick, 1981; Tate et al., 1990; Tibbets et al., 1989).

Given the critical functions of A-kinase in the cardiac cycle, changes in this system in response to the challenge of exercise training are important in the overall understanding of myocardial cellular metabolism.

Studies on the effects of exercise training on A-kinase activity were undertaken at the Kiev Institute of Physical Culture and Research Institute of Biochemistry, Academy of Science, Ukraine. It was found that A-kinase activity of cardiac muscle increases with exercise training in a dose-response manner, varying with the duration and perhaps the intensity of training (Kalinski & Kurski, 1983) (Table 4.2). The observed change in A-kinase activity most likely results from a change in its regulation by cAMP. The changes in kinetic parameters suggest that one of the training-induced causes of increased A-kinase activity may be an alteration of enzyme-substrate interactions (see Tables 4.3 and 4.4, Figures 4.3 and 4.4).

Because A-kinase is a cAMP-dependent enzyme and is regulated by catecholamines via AC, it is interesting to discuss the changes in the properties of A-kinase in relation to changes in [cAMP] and [catecholamine] as well as the

Table 4.2 Effect of Different Durations of Training on Activity of A-Kinase of Cardiac Muscle of Rats (n = 4)

	Activity nmol $^{32}P \cdot$ mg protein$^{-1} \cdot$ min^{-1}	
	Without cAMP	With cAMP
Group 1 (control)	0.21 ± 0.02	0.69 ± 0.07
Group 2 (trained 1.5 mo)	0.20 ± 0.02	1.66 ± 0.18[a]
Group 3 (trained 3.0 mo)	0.44 ± 0.04[a,b]	3.16 ± 0.29[a,b]

Note. Values are means ± SD.
[a]Indicates significant difference ($p < 0.05$) for trained versus control.
[b]Indicates significant difference ($p < 0.05$) for group 2 versus group 3.
The data are from Kalinski and Kurski (1983).

Table 4.3 Kinetic Characteristics (K_M for ATP and V_{max}) of A-Kinase of Cardiac Muscle of Trained Versus Untrained Rats (n = 4)

	K_M 10^{-5} M	V_{max} nmol $^{32}P \cdot$ min$^{-1} \cdot$ mg protein^{-1}
Group 1 (control)	1.05 ± 0.1	0.81 ± 0.07
Group 2 (trained 1.5 mo)	2.8 ± 0.4[a]	2.27 ± 0.25[a]
Group 3 (trained 3.0 mo)	0.50 ± 0.05[a,b]	3.85 ± 0.33[a,b]

Note. Values are means ± SD.
[a]Indicates significant difference ($p < 0.05$) for trained versus control.
[b]Indicates significant difference ($p < 0.05$) for group 2 versus group 3.
The data are from Kalinski and Kurski (1983).

Table 4.4 Kinetic Characteristics (K_M for Histone H2b) of A-Kinase of Cardiac Muscle of Trained Versus Untrained Rats (n = 4)

	K_M, · 10^{-5}m
Group 1 (control)	0.10 ± 0.01
Group 2 (trained 1.5 mo)	0.28 ± 0.03[a]
Group 3 (trained 3.0 mo)	0.50 ± 0.04[a,b]

Note. Values are means ± SD.

[a]Indicates significant difference ($p < 0.05$) for trained versus control.

[b]Indicates significant difference ($p < 0.05$) for group 2 versus group 3.

The data are from Kalinski and Kurski (1983).

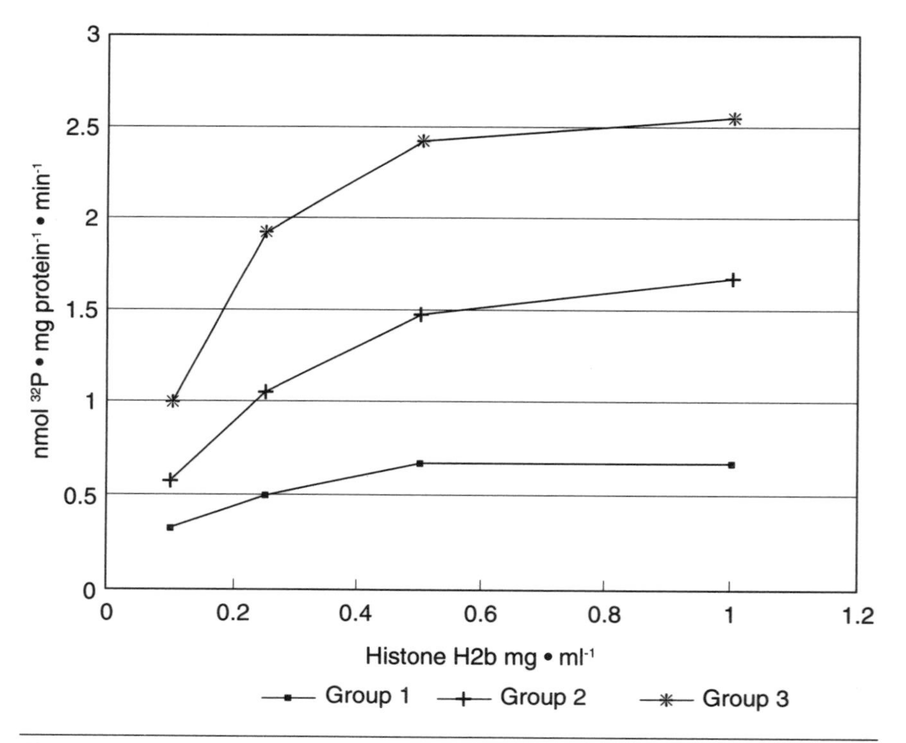

Figure 4.3 Activity of A-kinase of rat cardiac muscle across different concentrations of histone H2b. n = 4. 1, untrained control; 2, exercise trained 1.5 months; 3, exercise trained 3.0 months. Data from Kalinski & Kurski, 1983.

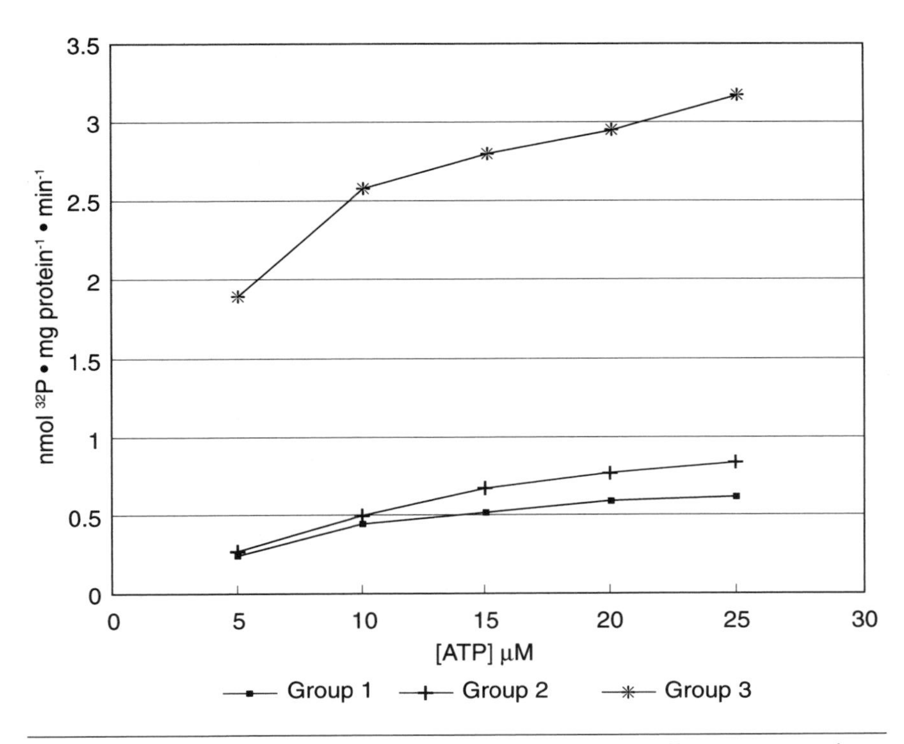

Figure 4.4 Activity of A-kinase of rat cardiac muscle across different concentrations of ATP. n = 4. 1, untrained control; 2, exercise trained 1.5 months; 3, exercise trained 3.0 months. Data from Kalinski & Kurski, 1983.

sensitivity of beta-adrenoreceptors to catecholamines and the activity of enzymes that control the levels of cAMP: AC and cAMP-PDE.

Resting myocardial [norepinephrine] has been shown to decrease in rats after three months of treadmill exercise training (DeShryver et al., 1969). However, increases in resting [NE] in the hearts of rats following swim training have also been found (Ostman & Sjostrand, 1972). The increase in NE turnover during exercise is attenuated in trained animals (Mazzeo, 1991).

It has also been shown that exercise training increases the sensitivity of beta-adrenergic receptors to catecholamines. This effect has been shown in cats (Hammond et al., 1988; Wyatt et al., 1978), rats (Takeda et al., 1985), and guinea pigs (Hammond et al., 1988) with both swim and run training. However, Mazzeo noted that conflicting results appear in the literature concerning the effects of exercise training on beta-adrenoreceptor number, density, and affinity (Mazzeo, 1991). The increases found in A-kinase activity might be related to either increased [NE] or increased sensitivity of beta-adrenergic receptors to catecholamine or perhaps a combination of both.

Acute bouts of exercise in untrained animals have been shown to increase [cAMP] and the activity of cAMP-PDE (Goldfarb et al., 1986; Goldfarb &

Kendrick, 1981; Palmer, 1988; Palmer et al., 1981). With respect to exercise-trained animals, Noakes et al. found no alteration in [cAMP] after 9 weeks of treadmill exercise training (Noakes, 1983), whereas Palmer et al. observed a higher myocardial [cAMP] in trained animals (Palmer et al., 1980). Dohm et al. (1976) found decreases in cardiac AC activity and no change in cardiac cAMP-PDE activity after 7 weeks of treadmill exercise training; these together might lead to decreased [cAMP]. The differences found in AC activity, and perhaps [cAMP], between Dohm et al. (1976) and Noakes et al. (1983) may be related to the training protocols. Dohm et al. (1976) used a higher exercise intensity and shorter duration than Noakes et al. (1983). Dohm et al. (1976) also had a shorter training period. As we stated previously, differences in intensity and duration of training have differing effects on cAMP metabolism. This is consistent with Kalinski's findings in that altering the duration and perhaps the intensity of exercise training altered the effect on A-kinase activity and also may depend on increased sensitivity of A-kinase to cAMP (Kalinski & Kurski, 1983) (Figure 4.5).

The striking increase in cardiac A-kinase activity following exercise training seems to be a response that may signal cellular adaptation. These

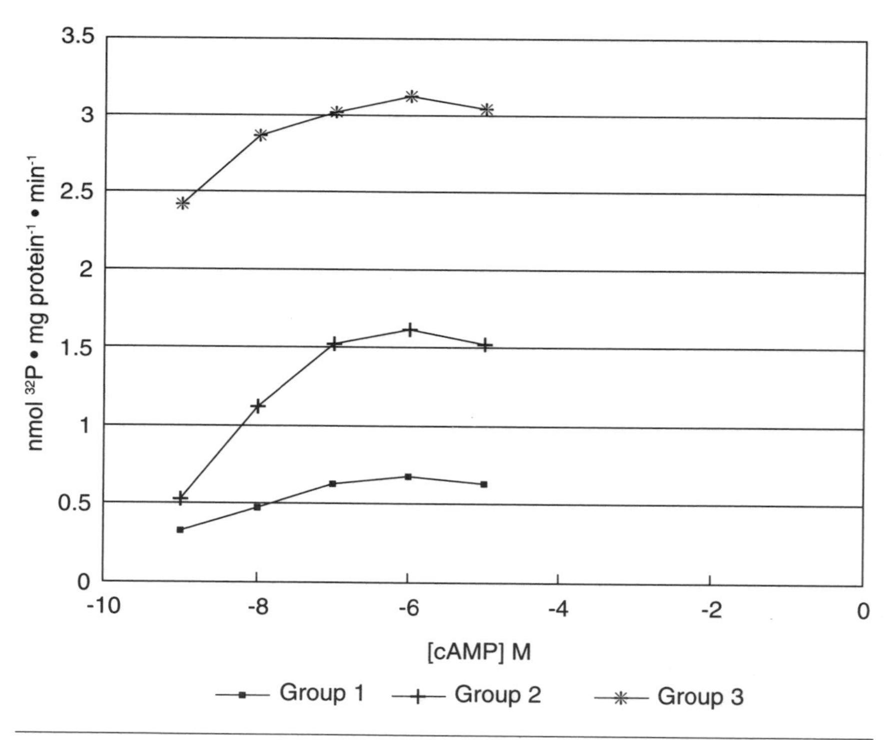

Figure 4.5 Activity of A-kinase of rat cardiac muscle across different concentrations of cAMP. $n = 4$. 1, untrained control; 2, exercise trained 1.5 months; 3, exercise trained 3.0 months. Data from Kalinski & Kurski, 1983.

changes in A-kinase activity would lead to enhanced phosphorylation of membrane proteins and the myofibrillar protein troponin I and enzymes. It has been demonstrated that endurance training increases the ability of A-kinase to phosphorylate such physiological substrates as the membrane proteins of SL and the SR of cardiac muscle (Kalinski et al., 1985; Kalinski & Zemtsova, 1982). In the presence of cAMP and A-kinase from the myocardium of control animals, phosphorylation of SL increased 3.3-fold as compared to a 4.9-fold increase seen in animals that underwent 6 weeks of exercise training (Kalinski et al., 1985). These data suggest that changes in cAMP-dependent phosphorylation may cause an adaptational increase in Ca^{++} flux across the SL and in the rate of Ca^{++} uptake by the SR of cardiac muscle. Tibbets et al. (1989) report such an increase following exercise training and Tate et al. (1990) have found an increased rate of Ca^{++} uptake by the SR of rats after exercise training. Exercise training also enhances the ability of A-kinase to regulate lysosomal myocardial cathepsin D release (Kalinski, 1984). These findings may relate to the adaptive improvements in contractile functioning seen after exercise training. Whether A-kinase loses its ability to regulate cathepsin D after detraining still needs to be clarified. It is interesting to note that after experimental myocardial infarction, cAMP-dependent phosphorylation of phospholamban decreased 1.6-fold (Antipenko et al., 1992); such a decrease may play a physiological role and lead to disturbance of contractile properties of cardiac muscle.

It is known that myocardial lipases are controlled by cAMP and A-kinase (Goldberg & Khoo, 1985; Palmer et al., 1986), and exercise has been demonstrated to affect the activity of lipoprotein lipase in rat heart (Borensztajn et al., 1975). Coupled with the training-induced alterations in A-kinase, this supports the concept that A-kinase is involved in training adaptations in the regulation of lipolysis.

Increased activity of cardiac A-kinase may also be responsible for the observed training-induced changes in properties of myofibrillar proteins troponin I and myosin isozymes, changes in contractility of cardiac muscle (Antipenko et al., 1992; Takeda et al., 1985; Winegrad et al., 1983), and accelerated protein synthesis (Watson et al., 1989; Xenophontos et al., 1989). High concentrations of A-kinase have been found in cardiac mitochondria (Schwoch et al., 1990), suggesting that enhanced activity of A-kinase after exercise training may be important in the adaptation of oxidative phosphorylation in the heart.

Possible effects of training-induced changes in A-kinase activity in the regulation of cardiac metabolism are shown in Figure 4.6. The enzyme A-kinase holds a pivotal role in the metabolic cascade of events involved in cardiac metabolism. It is apparent that increases in A-kinase activity lead to changes in numerous anabolic and catabolic processes that increase the ability of the heart to meet the demands of exercise. The enhanced cAMP-dependent phosphorylation

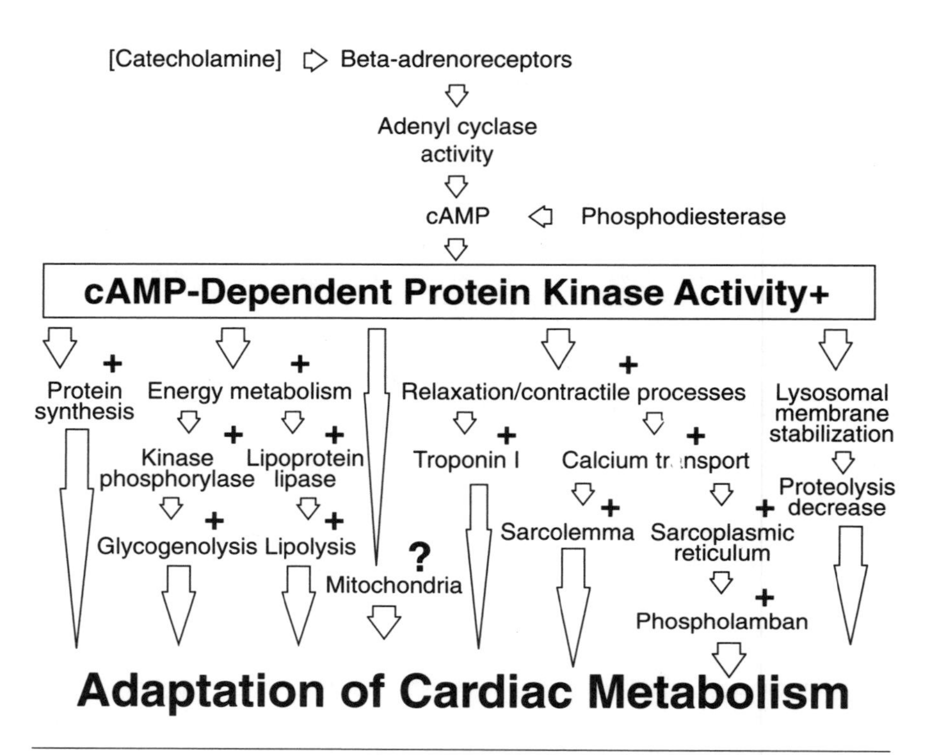

Figure 4.6 Role of cAMP-dependent phosphorylation in regulation of cardiac metabolism in adaptation to exercise training. + indicates exercise training stimulates.

of proteins involved in glycogenolysis, lipolysis, protein synthesis and degradation, and the transport of Ca^{++} all lead to major adaptations in metabolic activity and myocardial contraction and relaxation of cardiac muscle during exercise training. These changes are dependent on the duration and intensity of exercise training.

Chapter 5

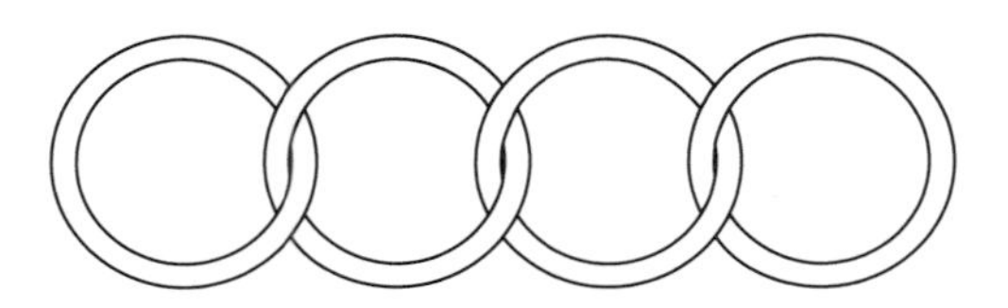

Effects of Prolonged Exercise and Training on Skeletal Muscle cAMP and cAMP-Dependent Protein Kinase

The consequences of both acute and chronic exercise on cAMP metabolism in skeletal muscle have received attention in scientific literature. However, almost no data have been presented in the English language on the influence of exercise on cAMP-dependent protein kinase in this tissue. This chapter introduces and assimilates data virtually unknown in the West on exercise and the overall cAMP system of skeletal muscle.

cAMP Metabolism

Changes in [cAMP] have been shown in both fast-twitch red and fast-twitch white fibers when rats were subjected to relatively short (<35 min) bouts of exercise at moderate and high intensities (Goldfarb et al., 1989; Sheldon et al., 1992). It has been shown that after a 2-week orientation period, an intensive run at 21m $\cdot$ min^{-1} on a 15% treadmill grade increased cAMP by 31% in the gastrocnemius/plantaris muscles of male Sprague-Dawley rats (Winder et al., 1989). Also, cAMP content of the white quadriceps muscle was significantly increased after 15 and 30 min of running (21 m $\cdot$ min^{-1} on a 15% treadmill grade) in fasted rats. In red quadriceps muscles of fasted rats, cAMP was significantly increased only after 30 min of running (Winder et al., 1991). Dohm et al. (1976) showed that AC activity in skeletal muscle was increased after exercise training.

Kalinski et al. (1980, 1984) examined the effects of prolonged exercise and physical training on cAMP, AC, and cAMP-PDE in skeletal muscle; they found that prolonged exercise (4-6 hr) led to decreases of 55% and 35%, respectively, in AC activity and in the [cAMP] in skeletal muscles of the untrained rats (Figure 5.1). They observed no significant changes in cAMP-PDE activity immediately after exercise; during recovery, [cAMP] and AC activity in untrained skeletal muscle were restored to control values at 24 and 36 hr, respectively (Kalinski et al., 1984). In this study it was also found that cAMP-PDE activity rose above the preexercise level from the 6th to the 48th hr (Figure 5.1).

Before exercise, trained animals show a significant increase in [cAMP] and AC activity compared to controls (Kalinski et al., 1980) (Table 5.1). The cAMP-PDE activity does not appear to differ between trained and untrained animals. After exercise, the [cAMP] in skeletal muscles is significantly reduced in trained animals, but to a lesser extent than in untrained animals (Kalinski et al., 1980) (Table 5.1). Activity of cAMP-PDE is significantly decreased as a result of exercise in trained rats (Kalinski et al., 1980) (Table 5.1).

Untrained animals showed a decrease in adrenal [epinephrine] after prolonged, mild exercise (Kalinski et al., 1969) (Figure 5.2). No change in adrenal [NE] has been observed. Active skeletal muscles show a significant decrease in [NE] after exercise. (Kalinski et al., 1969) (Figure 5.3).

Buckenmeyer et al. (1990) have shown no changes in AC activity in skeletal muscle of untrained rats that exercised for only 20 min. The work rate in this study was similar to that used by Kalinski et al. (1980). Therefore, it is apparent that the alterations in cAMP metabolism in skeletal muscle are time dependent.

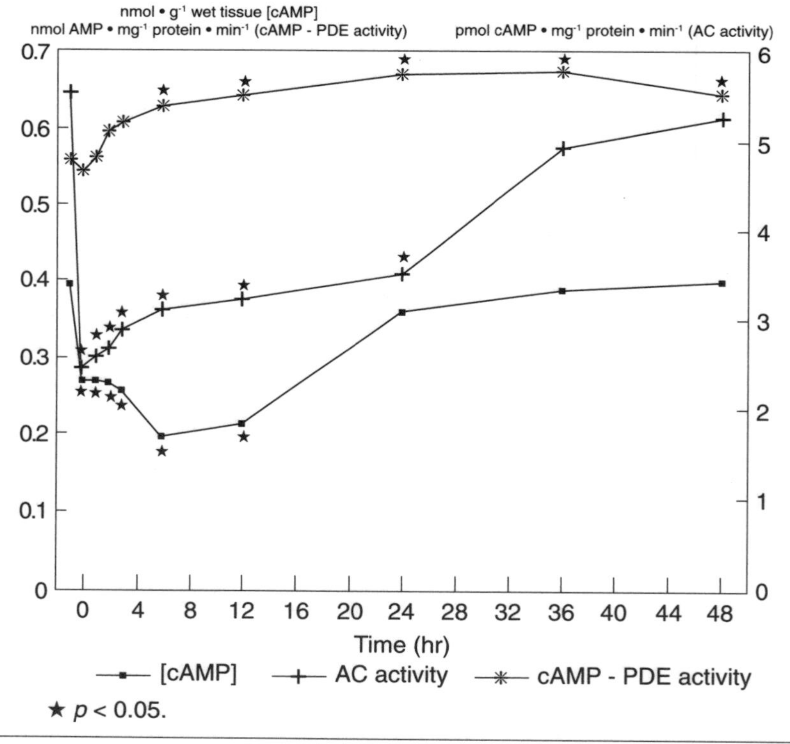

Figure 5.1 Alterations in cAMP metabolism in skeletal muscles of untrained rats following exercise. Data from Kalinski et al., 1984. Key: [cAMP], nmol · g^{-1} wet tissue; activity of AC, pmol cAMP · mg^{-1} protein · min^{-1}; cAMP-PDE, nmol AMP · mg $protein^{-1}$ · min^{-1}.

Table 5.1 [cAMP] and Activities of AC and cAMP-PDE in Skeletal Muscles of Untrained and Exercise-Trained Rats

	Untrained	Trained	
	Pre-ex	Pre-ex	Post-ex
n	6	6	6
[cAMP] pmol · g^{-1} wet tissue	970 ± 55	1286 ± 59[a]	1012 ± 70[b]
AC activity pmol cAMP · mg $protein^{-1}$ · min^{-1}	9 ± 1	15 ± 2[a]	6 ± 1[a,b]
cAMP-PDE pmol AMP · mg $protein^{-1}$ · min^{-1}	816 ± 80	785 ± 17	612 ± 20[a,b]

Note. Values are means ± SE.

[a]Comparisons of pre-ex or post-ex trained to pre-ex untrained.

[b]Comparisons of post-ex to pre-ex trained. Superscripts [a] and [b] indicate significant differences ($p < 0.05$).
The data are from Kalinski et al. (1980).

Figure 5.1 represents a sequential series of cellular events in the working muscles in which the steep decline in AC activity at the end of exercise was paralleled by a similar decline in the [cAMP]. These results are in contrast to those of Palmer (1988), who found a modest increase in [cAMP] of fast-twitch white fibers of rat vastus lateralis muscle but no significant changes in fast-twitch red and slow-twitch red fibers of the soleus muscle. The differences in results may be explained by the different modes of exercise used. One experimental protocol (Kalinski et al., 1984) used treadmill exercise, which may place greater metabolic demands (Booth & Thomason, 1991) on the rat's hamstring muscles than are placed on the soleus or vastus lateralis muscles of rats during a swimming protocol. In addition, the duration of the exercise in one study (Kalinski et al., 1984) was three times longer; this could have accounted for the steep declines in both AC activity and [cAMP]. Others have shown increases in cAMP content of all fiber types after short bouts of exercise (i.e., 10-30 min) (Goldfarb et al., 1989). As discussed earlier, these apparent contradictions are probably related to differences in exercise mode, duration, and intensity and differences in the muscles studied.

The postexercise period in untrained animals is characterized by no change in [cAMP] during the first 3 hr of recovery, followed by a decline at 6 hr of recovery (Kalinski et al., 1984) (Figure 4.1). This may be explained by the rise in cAMP-PDE. A steady state in [cAMP] between 6 and 12 hr may be explained by an equilibrium in its synthesis and degradation rates as evidenced by very small increments in both AC and cAMP-PDE activities, respectively. The rise in [cAMP] at the 24th hr postexercise was associated with an increase in AC

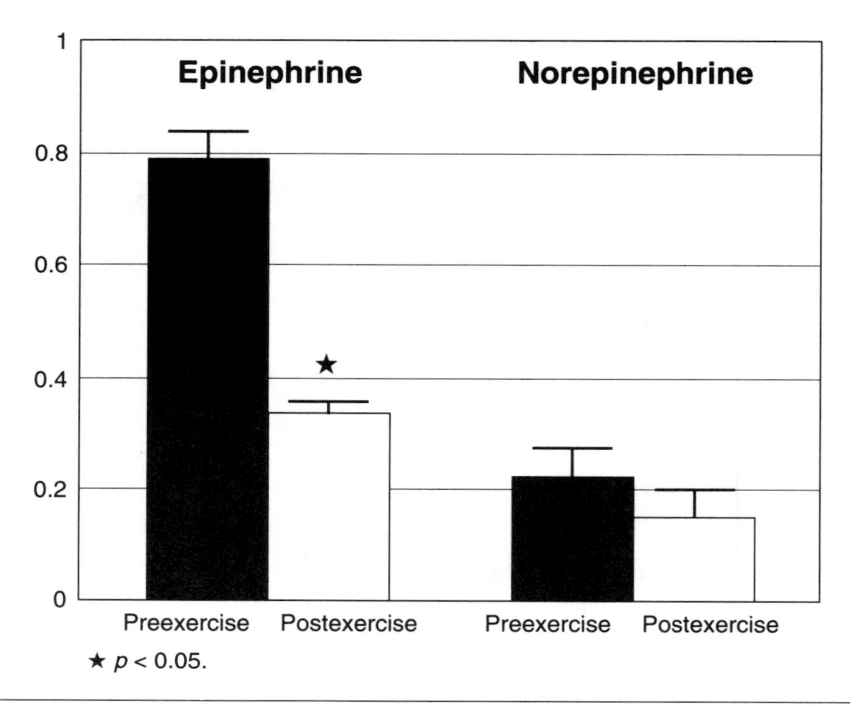

Figure 5.2 Effects of a single bout of prolonged, mild exercise on the [epinephrine] and [norepinephrine] in $\mu g \cdot mg^{-1}$ wet tissue of adrenal glands of untrained rats before and after exercise. Data from Kalinski et al., 1969.

activity. It is possible, then, that the rate of synthesis of [cAMP] was enhanced by the increased activity of AC, which exceeded the degradation rate of cAMP by cAMP-PDE.

After a single bout of exercise lasting 6 hr, the decline in [cAMP] in the trained animals was 21% as compared to 35% in the untrained rats (Kalinski et al., 1980, 1984) (Table 5.1 and Figure 5.1, respectively). The smaller decrease in [cAMP] was related to the higher [cAMP] present initially in the trained animals versus the two groups of untrained animals (Table 5.1 and Figure 4.1). Similar decreases in AC activity in the trained and untrained animals (Table 5.1 and Figure 5.1) were observed.

Exercise training provided the stimulus for skeletal muscle to increase its AC activity and synthesis of cAMP, resulting in an increase in [cAMP] before exercise as compared to levels in the untrained rats (Table 5.1). The increase of 67% in AC activity before exercise in the trained animals observed by Kalinski et al. (1980) is similar in direction to the change reported by Buckenmeyer et al. (1990) and Dohm et al. (1976).

Trained animals show significantly lower cAMP-PDE activities after a single bout of exercise than untrained (Kalinski et al., 1980, 1984) (Table 5.1 and Figure

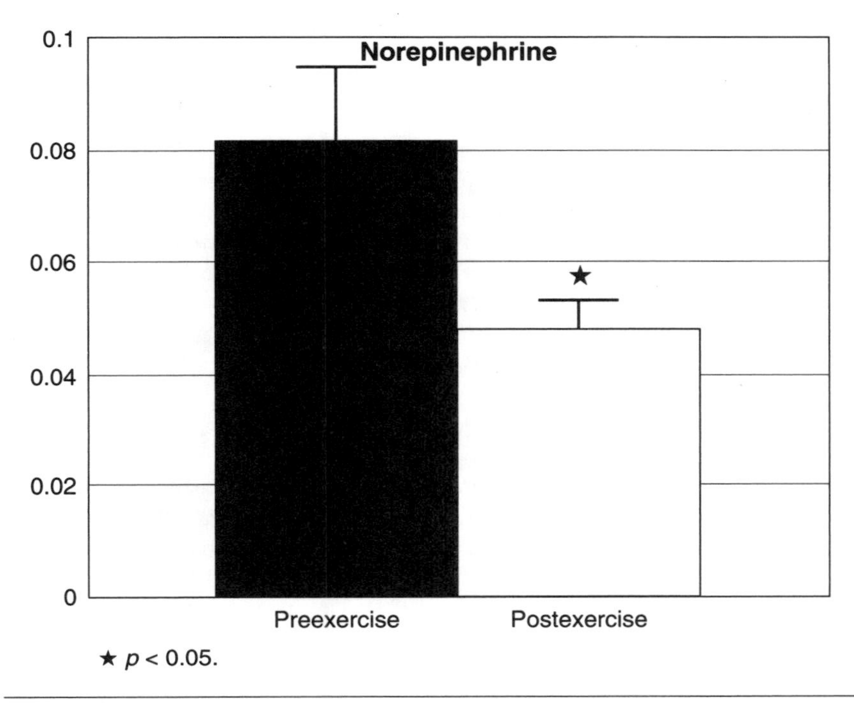

Figure 5.3 Effects of a single bout of prolonged, mild exercise on the [norepinephrine] in μg · g^{-1} wet tissue of skeletal muscles of untrained rats before and after exercise. Data from Kalinski et al., 1969.

5.1). These findings are consistent with the smaller decrease in [cAMP] after a single bout of exercise (Table 5.1). That is, lower cAMP-PDE activity levels preserved [cAMP] by decreasing the degradation rate of cAMP in trained skeletal muscle.

Winder and Duan (1992) have shown that the increases in cAMP and acceleration of glycolysis in less active white quadriceps muscle of fasted rats after 30 min of running at 21m · min^{-1} up a 15% grade are dependent on the rise in plasma epinephrine. The large decline in [epinephrine] in the adrenal glands after untrained animals exercise to exhaustion could be related to an enhanced rate of secretion. This would be coupled with a decrease in the rate of synthesis of epinephrine by the adrenal medulla (Malosheva & Matlina, 1973). The recognized, lipolytic effect of epinephrine in skeletal muscle during prolonged work is in part responsible for providing the energy for the active leg muscles of rats during exercise. Epinephrine has been shown to have profound effects on lipoprotein lipase activity in skeletal muscle, with the greatest increase occurring in the white fiber population of the vastus lateralis muscle of the rat (Miller et al., 1989). Possibly the degradation rate of NE is increased in the active muscles of untrained rats after exercise (Gorokhov, 1969). Palmer showed that adrenalectomy inhibited the exercise-induced increase in cardiac cAMP content in rats after 60 min of treadmill running (Palmer, 1988). The administration of propranolol had a similar effect (Palmer, 1988). Also, it has been shown that adrenodemedulation prevented the exercise-induced increase in cAMP in red and white quadriceps muscle of fasted rats (Winder & Duan, 1992). The findings of Kalinski et al. (1969) are in agreement with these data and suggest that the exercise-induced decrease in [cAMP] in skeletal muscle, as in cardiac muscle, may be mediated by a decrease of [epinephrine] in adrenal glands (Figure 5.2) and NE in skeletal muscle after 6 hr of running (Figure 5.3).

Data from Sheldon et al. (1992) show that in untrained rats after short term acute treadmill exercise, skeletal muscle cAMP levels are not associated with initiation of mitochondrial biogenesis for at least 4 hr postexercise.

cAMP-Dependent Protein Kinase

It is known that the effects of cAMP in skeletal muscle are mediated by cAMP-dependent protein kinase. Studies have found that physical training increased the ability of A-kinase to phosphorylate kinase phosphorylase b of skeletal muscle (Kalinski et al., 1982) and enhanced the ability to regulate lysosomal cathepsin D release in skeletal muscle (Kalinski, 1984). The increase in [cAMP] and AC activity may lead to changes in the activity of skeletal muscle A-kinase during acute exercise or exercise training.

As previously discussed, it has been shown that acute and chronic exercise induce significant changes in cAMP metabolism of skeletal muscle. The focus of this section will be on how exercise and exercise training affect the properties of A-kinase of skeletal muscle.

When skeletal muscle from rats was studied by chromatography on DEAE-cellulose, two isoforms of A-kinase could be isolated (Corbin et al., 1977). It is interesting to note that no differences in the number of isoforms of A-kinase in untrained skeletal muscle are found at rest or after exercise. However, the same exercise protocol causes an increase in the activity of A-kinase in its first and second isoenzymes as compared to rest (Figure 5.4).

Maximum activity of both isoenzymes of A-kinase of untrained rats is at a [cAMP] = 10^{-6} M (Kurski et al., 1978) (Figure 5.5). Table 5.2 indicates that both isoenzymes of A-kinase in skeletal muscles had very low activities without cAMP at rest or during exercise. However, when the isoenzymes were incubated with cAMP after an exercise training period of 1.5 months, the activities of isoenzymes I and II increased above resting values by 18% and 25%, respectively (Kalinski et al., 1981). No significant differences were observed between rest and exercise in the trained animals for either isoenzyme.

The kinetics of A-kinase were studied through use of substrates such as ATP and histone H2b (Kalinski et al., 1981) (Table 5.3). It can be seen that the K_m for ATP of both isoenzymes of A-kinase of skeletal muscles did not change. The V_{max} of A-kinase of skeletal muscles increased at rest after 1.5 months of

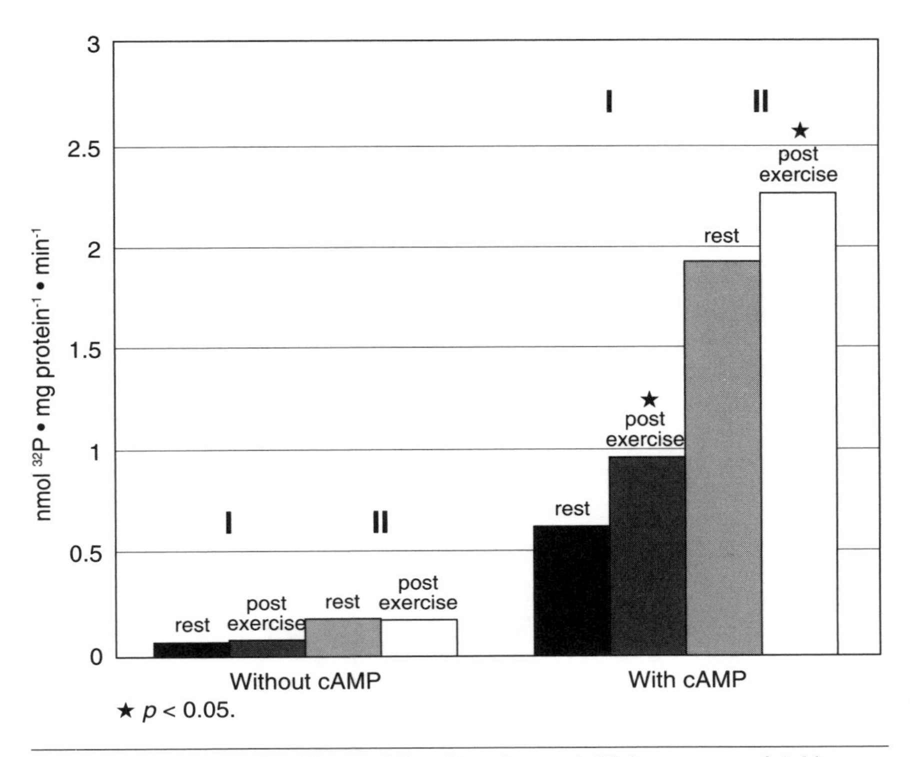

Figure 5.4 Activity of A-kinase of first (I) and second (II) isoenzymes of A-kinase in skeletal muscle of untrained rats at rest and after long-duration (5-6 h) exercise. Data from Kurski et al., 1978.

training (Table 5.3). At rest, the K_m for histone H2b of isoenzymes I and II of A-kinase of skeletal muscles increased and decreased, respectively, after 1.5 months of endurance exercise training (Kalinski et al., 1981) (Table 5.4). V_{max} was increased only for isoenzyme I of A-kinase of skeletal muscles after training (Table 5.4).

Activation of isoenzymes I and II of A-kinase by cAMP for trained skeletal muscles at rest is greater over a range of different concentrations of cAMP (Kalinski et al., 1981) (Figure 5.6). The optimal [cAMP] for A-kinase activity was at 10^{-6} M and is similar in both the trained and untrained states. The optimal pH for activity for both isoenzymes of A-kinase of skeletal muscles is unchanged after exercise training.

The activity of A-kinase at different [ATP] is shown in Figure 5.7. A-kinase activity increases in a curvilinear manner with [ATP] until it peaks at an [ATP] that is nearly identical for isoform I and II (Kalinski et al., 1981). The relationship of A-kinase activity and [H2b] is shown in Figure 5.8. After 1.5 months of training, A-kinase activity increases curvilinearly with increasing [histone H2b] for both isoenzymes of A-kinase of skeletal muscles (Kalinski et al., 1981) (Figure 5.8).

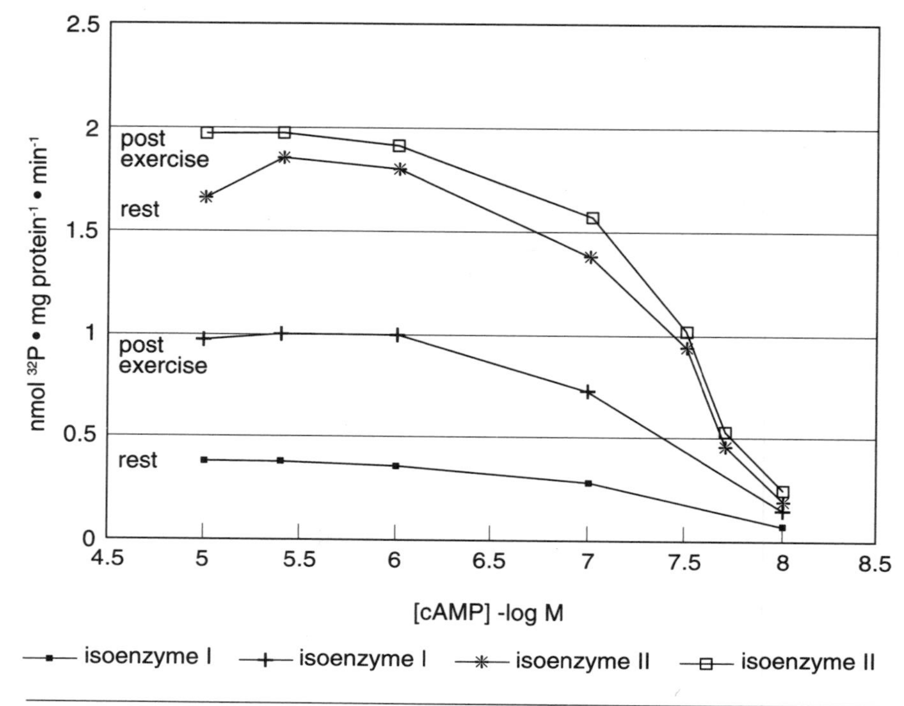

Figure 5.5 Activity of first (I) and second (II) isoenzymes of A-kinase at different concentrations of cAMP of skeletal muscle of untrained animals at rest and after acute exercise. Data from Kurski et al., 1978.

An analysis of the A-kinase activity in two isoforms of the enzyme in the skeletal muscles of untrained rats revealed increases of 35% and 15% in the A-kinase activities in isoenzymes I and II, respectively, after 5-6 hr of running (Kurski et al., 1978) (Figure 5.4). Isoform I of A-kinase is more sensitive to hormonal control (Corbin et al., 1977). This could explain the greater relative increase in isoform I after exercise (Figure 5.4).

It is both interesting and perplexing that despite the significant decreases of [NE] (Kalinski et al., 1969) and [cAMP] (Kalinski et al., 1984) in skeletal muscle of untrained rats after the same exercise protocol, the activity of A-kinase was increased. This unexpected result may be explained by an inhibition of a thermostable inhibitor of A-kinase. It has been shown that A-kinase activity can be inhibited in vitro in both untrained and trained skeletal muscle at rest and after similar acute exercise in the trained rat (Zemtsova et al., 1981). It has been shown in vitro that when isoform II of A-kinase was autophosphorylated, A-kinase activity increased twofold at rest and after exercise. Therefore, it is possible that the autophosphorylation mechanism was at least in part responsible for the increase in A-kinase activity.

Table 5.2 Activity of First and Second Isoenzymes of A-Kinase in Skeletal Muscles of Untrained and Trained Rats at Rest

	Isoenzyme I Activity – nmol $^{32}P \cdot mg\ protein^{-1} \cdot min^{-1}$	
	Without cAMP	With cAMP
Group 1 (untrained, rest)	0.14 ± 0.03	1.31 ± 0.04
Group 2 (trained, rest)	0.18 ± 0.01	1.60 ± 0.11[a]
	Isoenzyme II Activity – nmol $^{32}P \cdot mg\ protein^{-1} \cdot min^{-1}$	
Group 1 (untrained, rest)	0.22 ± 0.10	2.51 ± 0.24
Group 2 (trained, rest)	0.24 ± 0.04	3.35 ± 0.13[a]

Note. Values are means ± SD.

[a]Indicates significant difference ($p < 0.05$) for trained versus untrained.

The data are from Kalinski et al. (1981).

Table 5.3 Kinetic Characteristics of A-Kinase of Skeletal Muscle of Trained Versus Untrained Rats

	Isoenzyme I	
	$K_{m,} \cdot 10^{-5}m$	V_{max}, nmol $^{32}P \cdot min^{-1} \cdot mg^{-1}$ protein
Group 1 (untrained, rest)	0.13 ± 0.05	2.5 ± 0.09
Group 2 (trained, rest)	0.15 ± 0.05	4.36 ± 0.12[a]
	Isoenzyme II	
Group 1 (untrained, rest)	0.30 ± 0.09	2.33 ± 0.36
Group 2 (trained, rest)	0.41 ± 0.05	3.89 ± 0.27[a]

Note. Values are means ± SD.

[a]Indicates significant difference ($p < 0.05$) for trained versus untrained.

The data are from Kalinski et al. (1981).

Table 5.4 Kinetic Characteristics of A-Kinase of Skeletal Muscles of Trained Versus Untrained Rats (n = 6).

	Isoenzyme I	
	K_m, · 10^{-5}m	V_{max}, nmol ^{32}P · min^{-1} · mg^{-1} protein
Group 1 (untrained, rest)	1.02 ± 0.32	2.42± 0.73
Group 2 (trained, rest)	2.20 ± 0.15[a b]	4.51 ± 0.47[b]
	Isoenzyme II	
Group 1 (untrained, rest)	1.20 ± 0.08	5.40 ± 0.21
Group 2 (trained, rest)	0.98 ± 0.03[b]	5.93 ± 0.43

Note. Values are means ± SD.

[a]Indicates significant difference ($p < 0.01$) for trained versus untrained.

[b]Indicates significant difference ($p < 0.05$) for trained versus untrained.

The data are from Kalinski et al. (1981).

Experimental data suggest that endurance training is responsible for the significant increases in A-kinase activities of both isoenzyme I and II of skeletal muscle (Kalinski et al., 1981) (Table 5.2). Endurance training increases both AC activity and [cAMP] in the skeletal muscles of rats after identical exercise training protocols (Kalinski et al., 1980) and also with differing exercise training protocols (Buckenmeyer et al., 1990; Dohm et al., 1976). It appears, then, that the increase in A-kinase activity of both isoenzymes is coupled to the elevations in AC activity and [cAMP] levels that occur with endurance exercise training. The elevation in AC activity and the resulting [cAMP] elevation occur in sequence and precede the activation of the enzyme A-kinase. One of the important net results of these biochemical interactions has to be facilitation of the release of energy from intramuscular stores of substrate for the exercising muscle during prolonged work.

It is of some interest to note the significant increases in the V_{max} after endurance exercise training of both isoenzymes of A-kinase for the substrate ATP without changes in their K_m (Kalinski et al., 1981) (Table 5.3).

The increases in V_{max} and K_m were not consistently observed for histone H2b, which served as another substrate, in either of the two isoenzymes of A-kinase in exercise-trained animals (Kalinski et al., 1981). Increases in V_{max} of histone H2b in isoenzyme I indicate that isoenzyme I was more responsive to an endurance training program than was isoform II. This difference may have been attributable to the more than twofold lower initial level of V_{max} in isoform

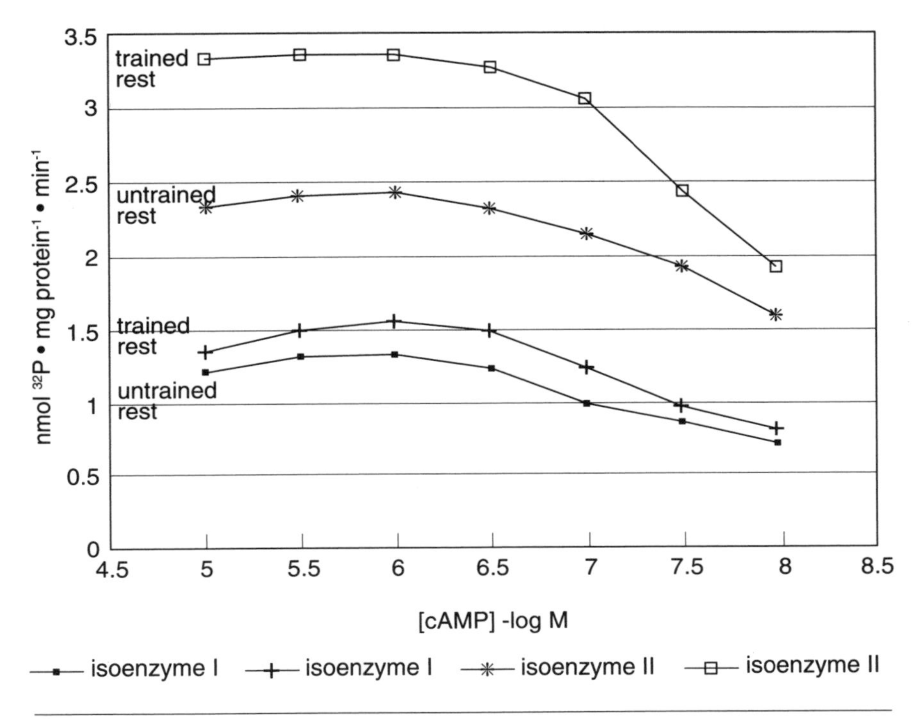

Figure 5.6 A-kinase activity of skeletal muscle in untrained and trained rats at rest and at different concentrations of cAMP (n = 6). Data from Kalinski et al., 1981.

I as compared to II. In addition, the more than twofold increase in K_m of isoform I is particularly interesting because it is opposite to the decrease observed in isoform II (Table 5.4). These significant differences may reflect a specificity of isoenzyme I to the exercise training that was not seen in isoform II. The net result was an increase in the capacity of isoenzyme I in endurance-trained skeletal muscle to phosphorylate histone H2b and perhaps skeletal muscle proteins also, which could enhance the metabolism of the trained muscle. The increase observed in the specific activities of isoenzymes I and II of A-kinase of skeletal muscles is a function of [cAMP] that regulates A-kinase activity. Additionally, the peak activities for both isoenzymes from both untrained and trained muscles occur at a [cAMP]-log M of 6 but the A-kinase activities are consistently higher in exercise-trained muscle (Kalinski et al., 1981) (Figure 5.6). In a similar manner, the activity of A-kinase increases curvilinearly with the [ATP] until it peaks at 5 μM [ATP] (Kalinski et al., 1981) (Figure 5.7). The activity of isoenzyme II of A-kinase is consistently higher than that of isoform I for both untrained and trained animals. These observations are consistent with the [cAMP] and A-kinase activity in Figure 5.6 and indicate that isoenzyme II in the trained rats is more

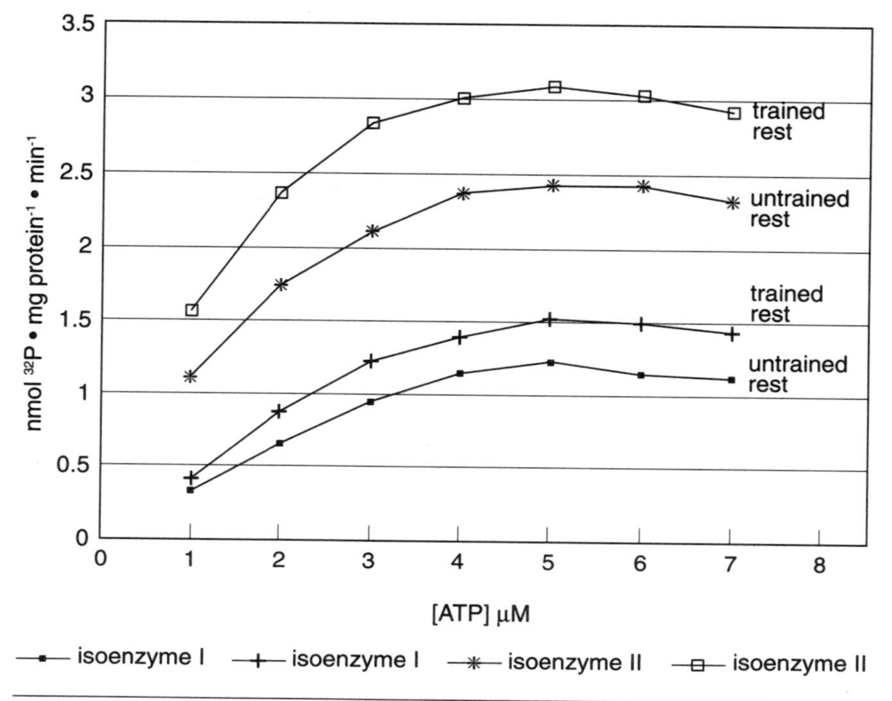

Figure 5.7 Activity of two isoenzymes of A-kinase of skeletal muscles for different concentrations of ATP in untrained (control) and trained animals at rest. Data from Kalinski et al., 1981.

sensitive than isoform I, and to a greater degree, in the trained rats than in the untrained at all [ATP]. It may well be that exercise training resulted in an increase in the sensitivity of A-kinase; this sensitivity is directly related to an increase in the capacity to do work by enhancement of [cAMP]. The increases in A-kinase activity are observed up to a peak [H2b] of 200 µg (Kalinski et al., 1981) (Figure 5.8). These increases are consistent for both isoenzymes I and II and the untrained and trained states. However, the differences in A-kinase activity are consistently greater in isoenzyme I at all [H2b]; in addition, the trained skeletal muscle exhibits a heightened sensitivity when compared to the untrained muscle. This increase in sensitivity of A-kinase in both isoforms of trained muscle may represent an increase in the affinity of the enzyme for its substrate. However, at that concentration, the activities of isoenzymes I and II of A-kinase most likely result from a change in A-kinase regulation by cAMP. We can assume that changes in kinetic parameters induced increases in enzyme activity by alteration of enzyme-substrate interactions (see Tables 5.3 and 5.4, Figure 5.8).

Exercise training may cause changes in tissue catecholamine levels and also turnover, synthesis, and activity of key enzymes related to catecholamine

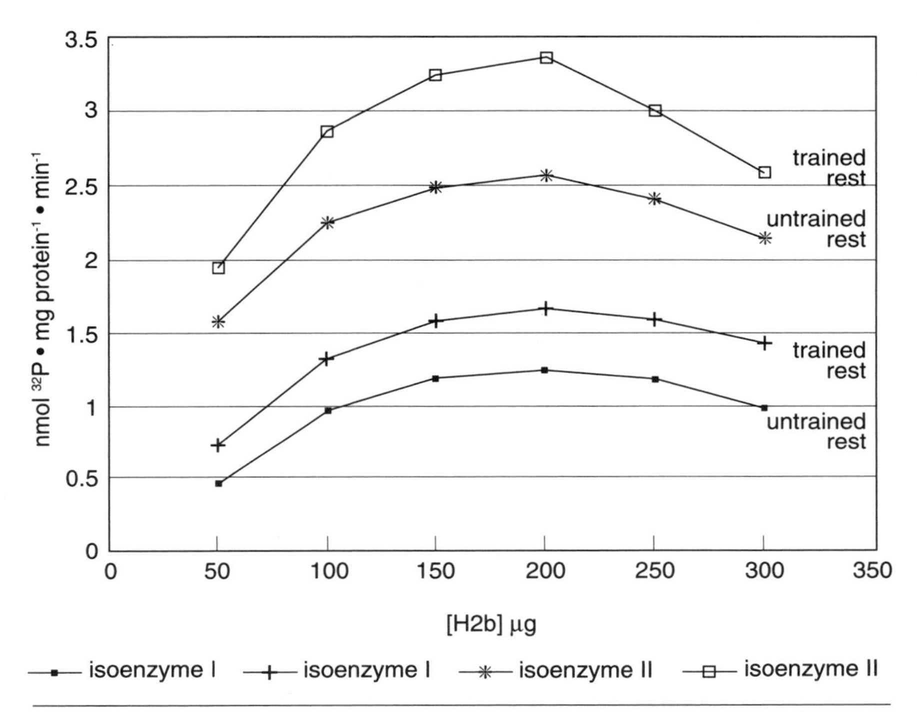

Figure 5.8 Activity of two isoenzymes of A-kinase of skeletal muscles for different [histone H2b] in untrained (control) and trained animals at rest. Data from Kalinski et al., 1981.

metabolism (Mazzeo, 1991). These events as well as increased beta-adrenergic receptor density in skeletal muscle as a result of aerobic training (Williams et al., 1984) may be related to increases of [cAMP] and A-kinase activity in the same tissue. Findings by Buckenmeyer et al. (1990) suggest that beta-adrenergic receptor density increased after endurance training in red vastus and in soleus muscles whereas in white vastus muscles beta-receptor density was not altered. Acute exercise did not alter beta-receptor density in trained rats (Buckenmeyer et al., 1990).

PART III

Summary

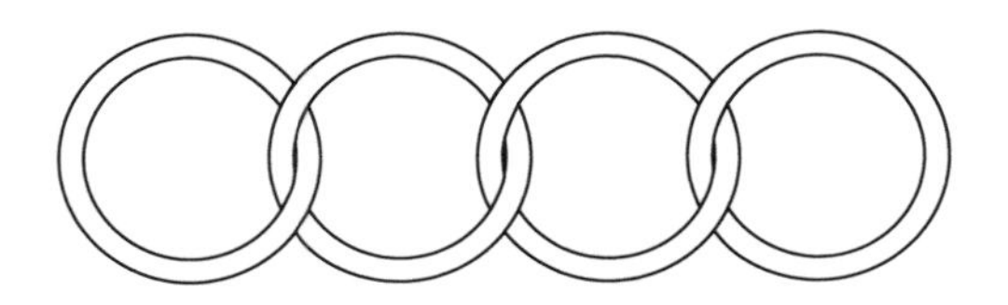

Chapter 6

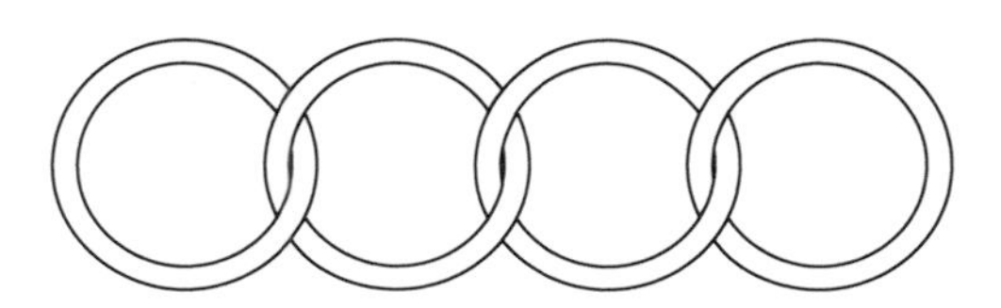

Present Understanding of the cAMP System in the Adaptations of Muscle to Exercise

The regulation of the myofibrillar complex, as well as regulation of free Ca^{++} levels in the cytosol of cardiac and skeletal muscle cells (which determines the functional activity), is controlled by different second messenger systems. According to the data obtained in the heart and in skeletal muscle, the main role in regulation in situ belongs to the cAMP-dependent protein kinase as compared with Ca^{++}-calmodulin-dependent and Ca^{++}-phospholipid-dependent kinases. As regards cGMP-dependent phosphorylation, its role in muscular activity is most pronounced in smooth muscle cells. This fact may be connected in particular with higher cGMP and G-kinase levels in smooth muscle cells as compared with cardiac and skeletal muscle (Collins et al., 1986; Franks et al., 1984). However, it should be emphasized that these conclusions cannot be regarded as final, since regulatory phosphorylation in the muscle cell is arousing the growing interest of investigators in the field of biochemistry of intracellular signals and new data can be expected to appear in the near future.

One should keep in mind that most of the findings described concerning the effect of second messengers on cardiac and skeletal muscle function were obtained in vitro. It is, however, obvious that the physiological significance of cAMP-dependent phosphorylation in myocytes is by no means a simple consequence of these findings. Thus, in extrapolating the data obtained in vitro to an in vivo situation, a number of factors should be taken into account, such as different contraction phases and compartmentalization of second messengers, phosphorylating, and dephosphorylating enzymes.

Compartmentalization of Second Messengers

It has been established that the cAMP level oscillates with the cardiac cycle during normal working conditions, with higher cAMP levels occurring during systole (Krause et al., 1989). The rise in cAMP during the systolic phase is accompanied by a transient activation of cytosolic A-kinase; assay of cGMP showed no oscillatory changes during the contraction-relaxation cycle (Krause et al., 1989). The fact that cAMP levels rise rapidly in the heart cycle suggests that early events occurring during membrane depolarization may stimulate AC or inhibit PDE. In turn, declines in cAMP during the second half of the heart cycle may be a result of Ca^{++} inhibition of AC or activation of PDE activity by this ion (Krause et al., 1989).

In this way, the normal oscillation of cAMP and A-kinase may be the product of neurohumoral regulation. This oscillation may lead to the specific changes in second messenger-dependent phosphorylation of regulatory membrane-bound and myofibrillar proteins. Thus, there is now evidence that phospholamban exists in a higher cAMP-dependent phosphorylation state during diastole than during systole (Krause, 1988).

Cyclic nucleotide and protein kinase compartmentalization in the myocytes also make it difficult to extrapolate the data obtained in vitro to the situation in the live cell. For instance, it is known that PGE induces the increase in cAMP levels in cardiac cells but, unlike isoprenaline, does not cause any positive inotropic effect (Tsien, 1977). In addition, under the influence of isoprenaline, cAMP levels and A-kinase activity increase both in the membrane compartments and in cytosol of cardiac cells, whereas under the influence of PGE these parameters increase only in the soluble fraction (Hayes & Brunton, 1982). Obviously, the difference in the effects on the myocardium between isoprenaline and PGE is not associated with the presence of different A-kinase isoforms in compartments. Thus, it is known that in the myocardium of many animal species, type II A-kinase is associated with the cell membrane. At the same time, isoprenaline and PGE have similar effects in the rat (80% type I A-kinase) and guinea pig (90% type II A-kinase) heart (Hayes & Brunton, 1982). The absence of a positive inotropic effect is ascribed to the lack of phosphorylation of a considerable number of regulatory target proteins in the cardiac cell during A-kinase activation (Hayes & Brunton, 1982). Similar effects on the myocardium are exerted by rolipram, a PDE inhibitor (Reeves et al., 1987). It may be suggested that these compounds induce the increase in cAMP levels and A-kinase activity in those compartments that do not contain target proteins. Rat heart perfusion with forskolin and isoprenaline leads to equal increases in cAMP levels in the heart; however, A-kinase activation and protein phosphorylation levels are significantly higher in the case of isoprenaline perfusion (England & Shahid, 1987). This provides evidence that forskolin increases cAMP levels in those compartments that are inaccessible for A-kinase.

The combination of the data thus obtained forms a basis for the hypothesis that cAMP levels increase in the submembrane compartments of the cardiac cell resulting in translocation of a dissociated catalytic A-kinase subunit from the insoluble into the soluble fraction (Murray et al., 1989). Such translocation can be observed in the presence of isoprenaline but not of PGE, indicating that this process is necessary for the development of a positive inotropic effect. However, this means that target proteins can ''choose'' between the catalytic subunit translocated into the cytosol and soluble A-kinase already present there. Thus, PGE induces a 65% activation of soluble A-kinase without increasing the contractility (Reeves et al., 1987). Hence, it can be concluded that a considerable portion of soluble A-kinase in the cardiac cell is spatially isolated from its potential substrates (Murray et al., 1989). This hypothesis is also supported by the finding by Aass et al. (1988) that the submembrane cAMP pool is the most important factor in the regulation of heart contractility. The data presented on the compartmentalization of second messengers, protein kinases, and protein phosphatases in muscle cells remind us repeatedly of the need to exercise caution in extrapolating results from an in vitro condition to one in situ.

Compartmentalization of second messengers, enzymes of second messenger metabolism, protein kinases, and protein phosphatases should also be taken into account when one is choosing therapeutic methods based on the modulation of strength and duration of intracellular signals. As is shown by practical experience, the use of compounds such as dibutyryl cAMP and other cyclic nucleotide analogues that penetrate through the cell membrane are by no means always efficient. Indeed, it has already become clear that in order to obtain a functional response, the increase or decrease in second messenger as well as the modulation of corresponding protein kinase activities is required not in the entire cell but in certain specific compartments. This effect can be achieved, for example, by specifically inhibiting or activating different cyclic nucleotide PDE forms localized in corresponding intracellular compartments.

Our knowledge of the basic cellular mechanisms helps us to better understand and appreciate the alterations in metabolism created by exercise. Exercise produces changes in hormone metabolism, but the intracellular mechanism of hormone actions during physical activity has not been sufficiently studied. It is universally accepted that some hormones regulate the function of cells via cyclic nucleotide-dependent protein kinases that catalyze the phosphorylation of proteins.

Significance of Changes in cAMP-Dependent Phosphorylation

The role of cAMP in regulating metabolism during exercise has not been fully elucidated. Untrained rats show decreases in cardiac [cAMP] and AC activity immediately following exercise. In the postexercise period, [cAMP] and AC activity peaked by the third hr and remained elevated during hr 6, 9, and 12 of recovery. The cAMP-PDE activity is significantly elevated at hr 12 and 24 of the postexercise period.

In trained animals, cardiac AC activity is elevated before exercise, and after exercise, [cAMP], AC, and cAMP-PDE activity increase. Therefore, endurance exercise training increases the capacity of the myocardial cell to synthesize cAMP during exercise. The augmentation of the cAMP system can lead to increased availability of energy for the cell and possibly enhance myocardial cell integrity and contractile function.

The function of A-kinase in regulating the adaptations to exercise training is not completely clear. Activity of myocardial A-kinase is increased in trained rats as compared to controls. The changes in A-kinase activity are dependent on the length of exercise training, with greater changes seen in the animals that underwent a longer training period. Changes in V_{max} and K_m of ATP and histone H2b also occurred and were also dependent on the duration of exercise training. The changes in A-kinase may have adaptive significance leading to increased glycolysis, lipolysis, oxidative phosphorylation, Ca^{++} transport in sarcoplasmic reticulum and sarcolemma, phosphorylation of troponin I, biosynthesis of proteins, and decreased cathepsin D release from lysosomes.

Our knowledge of cAMP and cAMP-dependent protein kinase metabolism of skeletal muscle is much more modest than that for cardiac muscle. After a prolonged bout of exercise, the [cAMP] in skeletal muscle is reduced in trained animals, but to a lesser extent than in untrained animals despite a decrease in AC activity. A reduction in PDE activity observed after exercise in the trained group helps to preserve the [cAMP]. Thus, endurance exercise training enhances the level of cAMP in skeletal muscle before an acute bout of exercise. This can provide an intracellular mechanism for increasing lipolysis and glycogenolysis in skeletal muscle during exercise. Before exercise, in trained rats, [cAMP] and AC activity is increased compared to that in the untrained animals.

A single bout of long-duration exercise increases the activities of both isozymes I and II of A-kinase in untrained animals in the presence of cAMP. Isozyme II activity is consistently higher than that of isoenzyme I at all cAMP concentrations at rest and after prolonged exercise. Endurance exercise training increases the activity of isozymes I and II in the presence of cAMP. At all concentrations of cAMP, ATP, and H2b, the activities of isozyme II are consistently higher in the trained than in the untrained state. In addition, the kinetic characteristic of V_{max} is consistently higher for both isozymes in resting trained rats compared to the untrained group. It may well be that exercise training heightens the sensitivity of both isozymes of A-kinase in skeletal muscle and may represent an increase in the affinity of the enzyme for its substrates.

The adenyl cyclase system is important in regulating the mitochondrial and sarcoplasmic enzymatic activities during glycogenolysis and lipolysis. Chronic electrical stimulation of skeletal muscle increases the content of cAMP and is related to an increase in mRNAs for mitochondrial proteins (Kraus et al., 1989). Lawrence and Salgiver (1964), using a tissue culture model and agents that enhanced [cAMP] in the cell of rat myotubes, observed increased levels of malate dehydrogenase and fumarase. There is also evidence of increases in the activities of other mitochondrial and sarcoplasmic enzymes, for example, NADH cytochrome c reductase and creatine kinase (Freerksen et al., 1986; Schutzle et al., 1984, respectively), which are also related to an increased content of cAMP in myotube cultures.

Increases in cAMP level may also be a potential factor in regulating a switch between fast and slow myosin isoforms in skeletal muscle (Kraus et al., 1989;

Zeman et al., 1988). However, the data from these two studies provide conflicting results.

Lipid mobilization is an essential process during prolonged exercise. It has been shown that intracellular alkaline lipoprotein lipase (LPL) of skeletal muscle is regulated by the classical cAMP cascade (Oscai et al., 1988). Palmer et al. (1990) have shown that when the cellular fraction of L8 myotube cultures were incubated with dibutyryl cAMP, the activity of TG lipase increased three- to five-fold. Peak activity occurred 16 hr after the addition of dibutyryl cAMP to the culture and returned to control levels 24 hours after its removal.

Oscai et al. (1988) interpret the blocking of intracellular LPL by a protein kinase inhibitor as evidence that LPL is regulated by cAMP-dependent protein kinase. Following this reasoning, increased activity of A-kinase of skeletal muscle may elevate the activity of intramuscular LPL and accelerate the mobilization of intracellular triglycerides. Thus, activity of intramuscular LPL may be enhanced by activation of A-kinase, leading to an increased release of free fatty acids into mitochondria that in turn results in increased mitochondrial oxidation of free fatty acids. This corresponds well with the concept of Holloszy and Coyle (1984) of glycogen sparing due to increased fat oxidation as a result of endurance training. However, key evidence is lacking; specifically, the demonstration that A-kinase isolated from skeletal muscle will phosphorylate intramuscular LPL-enzyme's protein and whether A-kinase from trained muscle will do this faster.

Exercise alters the rate of protein degradation both in red and white muscles (Dohm et al., 1980). Lysosomes of skeletal muscle contain acidic hydrolase-proteolytic enzymes that play a significant role in the intramuscular degradation of proteins. It has been demonstrated that acute exercise to exhaustion enhances the activity of cathepsin D and other lysosomal enzymes in red, white, and mixed muscle types (Vihko, Salminen, & Rantamaki, 1978). Interestingly, exercise training (running 1.5 hr for several weeks at 17.6 m $\cdot$ min^{-1}) increases activity of b-glucuronidase and does not change the activities of other acid hydrolases in red and white muscles of mice (Vihko, Hirsimaki, & Ajaman, 1978). It has been shown that increased intracellular [cAMP] decreases the release of cathepsin D and other proteolytic enzymes from lysosomes. The decreased release of proteolytic enzymes (including cathepsin D) is believed to be due to phosphorylation of lysosomal membranes mediated by A-kinase (Korovkin, 1982). After exercise training the ability of A-kinase to restrict lysosomal cathepsin D release in rat skeletal muscle is enhanced (Kalinski, 1984).

Changes in the activity of A-kinase also affect enzymes involved in glucose metabolism. Activation of phosphorylase kinase is caused by A-kinase. In turn, phosphorylase *b* kinase converts phosphorylase *b* (inactive) to phosphorylase *a* (active), which increases the rate of glycogen breakdown. A positive correlation between phosphorylase *b* kinase activity and the degree of A-kinase-dependent phosphorylation of this enzyme in skeletal muscle has been observed (Kalinski et al., 1982). An increased phosphorylase *b* kinase activity was observed to coincide with an increased deposition of ^{32}P in purified phosphorylase *b* kinase incubated with A-kinase from the skeletal muscle of trained versus untrained rats (Kalinski et al., 1982). Thus the increased activity of A-kinase that occurs with training probably plays an important role in the acceleration of energy mobilization from skeletal muscle glycogen.

Exercise training may cause a change in tissue catecholamine concentrations and in the turnover, synthesis, and activities of key enzymes related to catecholamine metabolism (Mazzeo, 1991).

Increases in beta-adrenergic receptor density in skeletal muscle as a result of aerobic training (Williams et al., 1984) may be related to the increases in [cAMP] and A-kinase activity. Buckenmeyer et al. (1990) noted differences in the effects of acute and chronic exercise on the beta-receptor density in different types of skeletal muscle fiber types. In red vastus and in soleus muscles, beta-receptor density increased whereas in white vastus muscles no differences were noted after endurance exercise training. However, with acute exercise, no changes in beta-receptor density were observed in the same muscles in untrained or trained rats (Buckenmeyer et al., 1990). Therefore, the enhancement of both [cAMP] and A-kinase activity of skeletal muscle resulting from endurance exercise training represented conditions that preceded, and were requisite for, the alterations and adaptations in cellular metabolism.

Some comparisons may be drawn between the responses of skeletal muscle and cardiac muscle to acute and chronic exercise. It has been shown that endurance exercise training increases the capacity of both types of muscle for cAMP before exercise by enhancing its synthesis. However, only cardiac muscle showed the capacity to increase the [cAMP] after exercise by increasing its synthesis via AC activity.

Many questions concerning the cAMP system remain unanswered. The roles of different second messengers and corresponding protein kinases (calmodulin-, phospholipid-, and cGMP-dependent) in the regulation of cardiac and skeletal muscle metabolism during exercise are not well understood. Also, the interaction of the second messenger system and Ca^{++}, lipid, and carbohydrate metabolism during exercise requires further investigation. The results of these investigations may be very promising because of the key role of phosphorylation in a myriad of metabolic processes.

References

Aass, H.; Skomedal, T.; Osnes, J. Increases of cyclic AMP in subcellular fractions of rat heart muscle after beta-adrenergic stimulation. J. Mol. Cell. Cardiol. 20:847-860; 1988.

Adelstein, R.S. Regulation of contractile proteins by phosphorylation. J. Clin. Invest. 72:1863-1866; 1983.

Akerman, K.E.O. Ca^{++} transport and cell activation. Med. Biol. 60:168-182; 1982.

Antipenko, A. Ye. Disturbances in protein phosphorylation and myocardial injury: causality or parallelism. Life Chem. Reports. 3/4:353-385; 1985.

Antipenko, A. Ye.; Lyzlova, S.N. Changes in the cAMP-dependent phosphorylation of myocardial sarcoplasmic reticulum under circulatory hypoxia. Biochem. (Moscow) 50:12-16; 1985.

Antipenko, A. Ye.; Kalinski, M.I.; Lyzlova, S.N. Myocardium metabolism regulation under different functional conditions. Ekaterynburg: Urals University Press; 1992.

Antipenko, A. Ye.; Stefanov, V.E.; Kraeva, L.N.; Lyzlova, S.N. Structural and functional modifications of phospholamban and regulation of calcium transport in the sarcoplasmic reticulum under conditions of the metabolic cardiac insufficiency. Ukrainian Biochem. J. 57:10-15; 1985.

Antipenko, A. Ye.; Sviderskaya, E.V.; Dizhe, G.P.; Krasovskaya, Ye. cAMP, calmodulin-dependent stimulation and resistance to proteolysis of the transport of Ca^{++} in sarcoplasmic reticulum of the heart. Biochem. (Moscow) 54:1652-1658; 1989.

Barany, M.; Barany, H.; Barron, J.T.; Kopp, S.J.; Doyle, D.D.; Hager, S.R.; Schlesinger, D.H.; Homa, F.; Sayers, S.T. Protein phosphorylation in live muscle. Protein phosphorylation. Cold Spring Harbor Conferences on Cell Proliferation 8:869-886; 1981.

Barany, M.; Barany, K. Phosphorylation of the myofibrillar proteins. Annu. Rev. Physiol. 42:275-292; 1980.

Barron, J.T.; Barany, M.; Barany, K.; Storti, R.V. Reversible phosphorylation and dephosphorylation of the 20,000-dalton light chain of myosin during the contraction-relaxation contraction cycle of arterial smooth muscle. J. Biol. Chem. 255:6238-6344; 1980.

Barry, W.H.; Smith, T.W. Movement of Ca^{++} across sarcolemma; effect of abrupt exposure to external Na^{+} concentration. J. Mol. Cell. Cardiol. 16:155-164; 1984.

Booth, F.W.; Thomason, D.B. Molecular and cellular adaptation of muscle in response to exercise: perspectives of various models. Physiol. Rev. 71(2):541-585; 1991.

Borensztajn, J.; Rone, M.S.; Babirak, S.P.; McGarr, J.A.; Oscai, J.A. Effect of exercise on lipoprotein lipase activity in rat heart and skeletal muscle. Am. J. Physiol. 229:394-397; 1975.

Bornet, E.P.; Entman, M.L.; Van Winkle, W.B.; Schwartz, A.; Lehotay, D.C.; Levey, G.S. Cyclic AMP modulation of calcium accumulation by sarcoplasmic reticulum from fast skeletal muscle. Biochim. Biophys. Acta. 468:188-193; 1977.

Briggs, F.N.; Lee, K.F.; Wechsler, A.W.; Jones, L.R. Phospholamban expressed in slow-twitch and chronically stimulated fast-twitch muscles minimally affects calcium affinity of sarcoplasmic reticulum Ca-ATPase. J. Biol. Chem. 267:26056-26061; 1992.

Brum, G.; Fitts, R.; Dizarro, G.F.; Rios, E. Voltage sensors of the frog skeletal muscle membrane require calcium to function in excitation-contraction coupling. J. Physiol. (London) 398:475-505; 1988.

Buckenmeyer, P.J.; Goldfarb, A.H.; Partilla, J.S.; Pinevro, M.A.; Dax, E.M. Endurance training, not acute exercise, differentially alters beta-receptor and cyclase in skeletal fiber types. Am. J. Physiol. 258(1):71-77; 1990.

Capony, J.P.; Rinaldi, M.L.; Guilleux, F.; Demaille, J.G. Isolation of cardiac membrane proteolipids by high pressure liquid chromatography: a comparison of reticular and sarcolemmal proteolipids, phospholamban and calciductin. Biochim. Biophys. Acta. 728:83-91; 1983.

Carafoli, E. Calmodulin in the regulation of calcium fluxes in cardiac sarcolemma. Adv. Myocardial. 56:97-101; 1985.

Carafoli, E. Intracellular calcium homeostasis. Annu. Rev. Biochem. 56:395-433; 1987.

Caroni, P.; Carafoli, E. An ATP-dependent, Ca^{++}-pumping system in dog heart sarcolemma. Nature. 238:765-767; 1980.

Caroni, P.; Carafoli, E. The regulation of the Na^{+}-Ca^{++} exchanger of heart sarcolemma. Eur. J. Biochem. 132:451-460; 1983.

Catterall, W.A. Excitation-contraction coupling in vertebrate skeletal muscle: A tale of two calcium channels. Cell. 64:871-874; 1991.

Cavadore, J.C.; De Peuch, C.G.; Walsh, M.P.; Vallet, B.; Molla, A.; Demaille, J.G. Calcium-calmodulin-dependent phosphorylation in the control of muscular contraction. Biochimie. 63:301-306; 1981.

Chacko, S.; Conti, M.A.; Adelstein, R.S. Effect of phosphorylation of smooth muscle myosin on actin activation and Ca^{++} regulation. Proc. Natl. Acad. Sci. U.S.A. 74:129-133; 1977.

Charles, L.; Bruce, J. Affinity labeling of calmodulin-binding components in canine cardiac sarcoplasmic reticulum. J. Biol. Chem. 257:15187-15191; 1982.

Chiesi, M.; Gasser, J. The regulation of the Ca^{++}-pumping activity of cardiac sarcoplasmic reticulum by calmodulin. In: Gerday, Ch.; Gilles, R.; Bolis, L., eds. Calcium and calcium binding proteins. Berlin, Heidelberg: Springer-Verlag; 1988: p. 220-226.

Chiesi, M.; Schwaller, R. Involvement of electrostatic phenomena in phospholamban induced stimulation of Ca^{++} uptake into cardiac sarcoplasmic reticulum. FEBS Lett. 244:241-244; 1989.

Cockroft, S. Inositol lipids, G-proteins and signal transduction. In: Gerday, Ch.; Gilles, R.; Bolis, L., eds. Calcium and calcium binding proteins. Berlin, Heidelberg: Springer-Verlag; 1988: p. 3-20.

Cohen, P. Signal integration at the level of protein kinases, protein phosphatases and their substrates. Trends Biochem. Sci. 17:408-413; 1992.

Collins, P.; Griffith, T.M.; Henderson, R.H.; Lewis, M.J. Endothelium-derived relaxing factor alters calcium fluxes in rabbit aorta: a cyclic guanosine monophosphate-mediated effect. J. Physiol. (London) 381:427-437; 1986.

Colyer, J.; Wang, J.H. Dependence of cardiac sarcoplasmic reticulum calcium pump activity on the phosphorylation status of phospholamban. J. Biol. Chem. 26:17486-17493; 1991.

Corbin, J.D.; Sugden, P.H.; Lincoln, T.M.; Keely, S.L. Compartmentalization of adenosine 3′, 5′-monophosphate and adenosine 3′, 5′-monophosphate dependent protein kinase in heart tissue. J. Biol. Chem. 252:3854-3861; 1977.

Costa, M.; Gerner, E.W.; Russel, D.M. Cell cycle-specific activity of type I and type II cyclic adenosine 3′, 5′ monophosphate-dependent protein kinases in Chinese hamster ovarian cells. J. Biol. Chem. 251:3312-3319; 1976.

Cuppoletti, J.; Thakkar, J.; Sperelakis, N.; Wahler, G. Cardiac sarcolemmal substrate of the cGMP-dependent protein kinase. Membr. Biochem. 7:135-142; 1988.

Curtis, B.M.; Catterall, W.A. Phosphorylation of the calcium antagonist receptor of the voltage-sensitive calcium channel by cAMP-dependent protein kinase. Proc. Natl. Acad. Sci. U.S.A. 28:2528-2532; 1985.

De Jongh, K.S.; Merrick, D.K.; Catterall, W.A. Subunits of purified calcium channels: a 212-kDa form of and partial amino acid sequence of a phosphorylation site of an independent subunit. Proc. Natl. Acad. Sci. U.S.A. 86:8585-8589; 1989.

DeSchryver, C.; Mortens-Strythagen, J.; Becsei, I.; Lammerant, J. Effect of training on heart and skeletal muscle catecholamine concentration in rats. Am. J. Physiol. 217:1589-1594; 1969.

Dixon, D.A.; Haynes, D.H. Kinetic characterization of the Ca^{++} pumping ATPase of cardiac sarcolemma in four states of activation. J. Biol. Chem. 264:13612-13622; 1989.

Dohm, G.L.; Kasperek, G.J.; Tapscott, E.B.; Beecher, G.R. Effect of exercise and on synthesis and degradation of muscle protein. Biochem. J. 188:255-262; 1980.

Dohm, G.L.; Pennington, S.N.; Barakat, H. Effect of exercise training on adenyl cyclase and phosphodiesterase in skeletal muscle, heart, and liver. Biochem. Med. 16:138-142; 1976.

Dolphin, A.C. Regulation of calcium channel activity by GTP binding proteins and second messengers. Biochim. Biophys. Acta. Mol. Cell. Res. 1091:68-80; 1991.

Ebashi, S. Ca and the contractile proteins. J. Mol. Cell. Cardiol. 16:129-136; 1984.

England, P.J. Studies on the phosphorylation of the inhibitory subunit of troponin during modification of contraction in perfused rat heart. Biochem. J. 160:295-304; 1976.

England, P.J. Protein phosphorylation in the regulation of muscle contraction. In: Cohen, P., ed. Recently discovered systems of enzyme regulation by reversible phosphorylation. Amsterdam: Elsevier; 1980: p. 153-173.

England, P.J. A comparison of the regulation of vascular and cardiac muscle. Biochem. Soc. Trans. 16:503-505; 1988.

England, P.J.; Krause, E.G. The effect of hypoxia on the phosphorylation of contractile and other proteins in perfused rat heart challenged by isoprenaline. Biomed. Biochim. Acta. 46:369-380; 1987.

England, P.J.; Mills, D.; Pask, H.T.; Jeacocke, S.A. The phosphorylation of cardiac contractile proteins. Abh. Acad. Wiss. DDR Abt. Math. Naturwiss. Techn. 1:29-34; 1984.

England, P.J.; Shahid, M. Effects of forskolin on contractile responses and protein phosphorylation in the isolated perfused rat heart. Biochem. J. 246:687-695; 1987.

Enyedi, A.; Farago, A.; Sarcadi, B.; Gardos, G. Cyclic AMP-dependent protein kinase and Ca-calmodulin stimulate the formation of pholyphosphoinositides in a sarcoplasmic reticulum preparation of rabbit heart. FEBS Lett. 176:235-238; 1984.

Fabiato, A.; Fabiato, F. Cyclic AMP-induced enhancement of calcium with no modification of the sensitivity of the myofilaments to calcium in skinned fibers from a fast skeletal muscle. Biochim. Biophys. Acta. 534:253-260; 1978.

Fabiato, A.; Fabiato, F. Calcium and cardiac excitation-contraction coupling. Annu. Rev. Physiol. 41:473-484; 1979.

Fischmeister, R.; Hartzell, H.C. Mechanism of action of acetylcholine on calcium current in single cells from frog ventricle, J. Physiol. (London) 376:183-202; 1986.

Fowler, C.; Huggins, J.; Hall, C.; Restall, C.J.; Chapman, D. The effect of calcium, temperature and phospholamban phosphorylation on the dynamics of the calcium-stimulated ATPase of canine cardiac sarcoplasmic reticulum. Biochim. Biophys. Acta. 980:348-356; 1989.

Franks, K.; Cooke, R.; Stull, J.T. Myosin phosphorylation decreases the ATPase activity of cardiac myofibrils. J. Mol. Cell. Cardiol. 16:597-604; 1984.

Freerksen, D.L.; Schroedl, N.A.; Johnson, G.V.W.; Hartzel, C.R. Increased aerobic glucose oxidation by cAMP in cultured regenerated skeletal myotubes. Am. J. Physiol. 250 (Cell Physiol. 19): C713-C719; 1986.

Galani-Kranias, E.; Bick, R.; Schwartz, A. Phosphorylation of a 100,000 dalton component and its relationship to calcium transport in sarcoplasmic reticulum from rabbit skeletal muscle. Biochim. Biophys. Acta. 628:438-450; 1980.

Gasser, J.; Paganetti, P.; Carafoli, E.; Chiesi, M. Heterogeneous distribution of calmodulin- and cAMP-dependent regulation of Ca^{++} uptake in cardiac sarcoplasmic reticulum subfractions. Eur. J. Biochem. 176:535-541; 1988.

Goldberg, D.I.; Khoo, J.C. Activation of myocardial neutral triglyceride lipase and neutral cholesterol esterase by cAMP-dependent protein kinase. J. Biol. Chem. 260:5879-5882; 1985.

Goldfarb, A.H.; Bruno, J.F.; Buckenmeyer, P.J. Intensity and duration effects of exercise on heart cAMP, phosphorylase, and glycogen. J. Appl. Physiol. 60(4):1268-1273; 1986.

Goldfarb, A.H.; Bruno, J.F.; Buckenmeyer, P.J. Intensity and duration of exercise effects on skeletal muscle cAMP, phosphorylase, and glycogen. J. Appl. Physiol. 66(1):190-194; 1989.

Goldfarb, A.H.; Kendrick, Z.V. Effect of an exercise run to exhaustion on cAMP in the rat heart. J. Appl. Physiol.: Respir. Environ. Exerc. Physiol. 51(6):1539-1542; 1981.

Gonzales-Serratos, H.; Hill, L.; Valle-Aguilera, R. Effects of catecholamines and cyclic AMP on excitation-contraction coupling in isolated skeletal muscle fibers of the frog. J. Physiol. (London) 315:267-282; 1981.

Gorochow, A.L. Action of muscular activity on catecholamine content in tissues of trained and untrained white rats. Sechenov. Physiol. J. U.S.S.R. 55:1411; 1969.

Gusev, N.B.; Dobrovolsky, A.B. Phosphorylation of troponin - facts, hypothesis and suppositions. Biochem. (Moscow). 41:1338-1341; 1976.

Hammond, H.K.; Ransnas, L.A.; Insel, P.A. Noncoordinate regulation of cardiac G_s protein and beta-adrenergic receptors by a physiological stimulus, chronic dynamic exercise. J. Clin. Invest. 82:2168-2171; 1988.

Hartweg, D.; Bafer, H. Studies on protein composition of cardiac sarcoplasmic reticulum membranes (SR). Naunyn Schmiedebergs Arch. Pharmacol. 322:25-30; 1983.

Hartzell, H.C.; Fischmeister, A. Opposite effects of cyclic GMP and cyclic AMP on Ca current in single heart cells. Nature. 323:273-276; 1986.

Hartzell, H.C.; Titus, L. Effects of cholinergic and adrenergic agonists on phosphorylation of a 165,000 dalton myofibrillar protein in intact cardiac muscle. J. Biol. Chem. 257:2111-2120; 1982.

Hawthorne, J.N.; Simmonds, S.H. Second messengers involved in the muscarinic control of heart: the role of the phosphoinositide response. Mol. Cell. Biochem. 89:187-189; 1989.

Hayes, J.S.; Brunton, L.L. Functional compartments in cyclic nucleotide action. J. Cyclic Nucl. Res. 8:1-16; 1982.

Hicks, M.J.; Chikegawa, M.; Katz, A.M. Mechanism by which cyclic adenosine 3′, 5′-monophosphate-dependent protein kinase stimulates calcium transport in cardiac sarcoplasmic reticulum. Circ. Res. 44:384-391; 1979.

Holloszy, J.O.; Coyle, E.F. Adaptations of skeletal muscle to endurance exercise their metabolic consequences. J. Appl. Physiol. 56:831-838; 1984.

Holroyde, M.J.; Howe, E.; Solaro, R.J. Modification of calcium requirements for activations of cardiac myofibrillar ATPase by cyclic AMP-dependent phosphorylation. Biochim. Biophys. Acta. 586:63-70; 1979.

Hosey, M.M.; Borsotto, M.; Lazdunski, M. Phosphorylation and dephosphorylation of dihydropyridine-sensitive voltage-dependent Ca^{++} channel in skeletal muscle membranes by cAMP- and Ca^{++}-dependent processes. Proc. Natl. Acad. Sci. U.S.A. 83:3733-3737; 1986.

Houslay, M.D. "Crosstalk": a pivotal role for protein kinase C in modulating relationships between signal transduction pathways. Eur. J. Biochem. 195:9-27; 1991.

Huggins, J.P.; Cook, E.A.; Piggott, J.R.; Mattinsley, T.J.; England, P.J. Phospholamban is a good substrate for cyclic GMP-dependent protein kinase in vitro, but not in intact, cardiac or smooth muscle. Biochem. J. 260:829-835; 1989.

Huggins, J.P.; England, P.J. Sarcolemmal phospholamban is phosphorylated in isolated rat hearts perfused with isoprenaline. FEBS Lett. 163:297-302; 1983.

Iwasa, J.; Hosey, M. Phosphorylation of cardiac sarcolemma proteins by the calcium-activated phospholipid-dependent protein kinase (protein kinase C). J. Biol. Chem. 259:534-540; 1984.

Iyer, R.B.; Koritz, S.B.; Kirchberger, M.A. A regulation of the level of phosphorylated phospholamban by inhibitor-1 in rat heart preparation in vitro. Mol. Cell. Endocrinol. 55:1-6; 1988.

Jakab, G.; Kranias, E.G. Phosphorylation and dephosphorylation of purified phospholamban and associated phosphatidylinositides. Biochem. 27:3799-3806; 1988.

Jakab, G.; Rapundalo, S.T.; Solaro, R.J.; Kranias, E.G. Phosphorylation of phospholipids in isolated guinea pig hearts stimulated with isoprenaline. Biochem. J. 251:189-194; 1988.

Jett, M.F.; Schworer, C.M.; Bass, M.; Soderling, T.R. Identification of membrane-bound calcium, calmodulin-dependent protein kinase II in canine heart. Arch. Biochem. Biophys. 255:354-360; 1987.

Jones, L.R.; Maddock, S.W.; Besh, H.R. Unmasking effect of alamethicin on cAMP-dependent protein kinase activity of cardiac sarcolemmal vesicles. J. Biol. Chem. 255:9971-9980; 1980.

Jones, L.R.; Maddock, S.W.; Hathaway, D.R. Membrane localization of myocardial type II cyclic AMP-dependent protein kinase activity. Biochim. Biophys. Acta. 641:242-253; 1981.

Jones, L.R.; Simmermann, H.K.; Wilson, W.W.; Gurd, F.R.; Wegener, A.D. Purification and characterization of phospholamban from canine cardiac sarcoplasmic reticulum. J. Biol. Chem. 260:7721-7730; 1985.

Jorgensen, A.O.; Jones, L.R. Localization of phospholamban in slow but not fast canine skeletal muscle fibers. J. Biol. Chem. 261:3775-3781; 1986.

Kalinski, M.I. Influence of 3′, 5′-AMP-dependent protein kinase on cathepsin D release from lysosome of skeletal muscles and heart of rats during physical exercise. Ukrainian Biochem. J. 56(4):408-413; 1984.

Kalinski, M.I., Gubsky, Yu. I.; Rudnitskaya, N.D.; Kurski, M.D. ATP-dependent transport of CA^{2+} in myocardial sarcoplasmic reticulum during adaptation to muscular activity. Questions of Medical Chemistry (Moscow). 4:31-34; 1989.

Kalinski, M.I.; Kondratyuk, T.P.; Kurski, M.D. cAMP-dependent phosphorylation phosphorylase "b" kinase of skeletal muscle of rats during physical exercise and training. Biochem. (Moscow). 47(12):1988-1992; 1982.

Kalinski, M.I.; Kononenko, V.J.; Roudenko, A.A. Influence of exercise and pharmacological agents on catecholamine metabolism. The II European Congress on Sports Medicine, Bucharest; 1969.

Kalinski, M.I.; Kotsuruba, V.N.; Chalmuradov, A.G.; Gubkina, N.I.; Balakleevski, A.A. Influence of immobilization of adrenaline on adenylate cyclase system of cardiac and skeletal muscle in recovery period after exercise. Ukrainian Biochem. J. 56(1):52-57; 1984.

Kalinski, M.I.; Kurski, M.D. Changes in some properties of soluble 3′, 5′-AMP-dependent protein kinase of myocardium during exercise training. Biochem. (Moscow). 48(8):1324-1328; 1983.

Kalinski, M.I.; Kurski, M.D.; Zemtsova, I.I.; Osipenko, A.A. Changes of some properties of 3′, 5′-AMP-dependent protein kinase of skeletal muscle of rats during endurance exercise training. Biochem. (Moscow). 46(1):120-125; 1981.

Kalinski, M.I.; Osipenko, A.A.; Kondratyuk, T.P.; Kurski, M.D. Characterization of protein kinase from myocardium and use for study of content of cAMP in tissue during exercise. Ukrainian Biochem. J. 49(3):99-103; 1977.

Kalinski, M.I.; Rudenko, A.A.; Kononenko, V. Ya. Effect of acute exercise and training on the catecholamine in different tissues and EKG of white rats. In: Viru, A.; Zimkin, N.; Yakovlev, N., eds. Endocrine mechanism of regulation of adaptation to muscular activity. Estonian S.S.R.: Tartu State University; 1969: p. 287-295.

Kalinski, M.I.; Tyutyunnik, W.R.; Stefanov, A.V.; Lishko, V.K. The influence of biogenic amines encapsulated in liposomes on energy metabolism during exercise. Sport Biochemistry: Proceedings of Leningrad International Symposium on Exercise Biochemistry, Leningrad: Research Institute of Physical Culture; 1990: p. 122-129.

Kalinski, M.I.; Vorobets, Z.D.; Fedotowa, E.U. cAMP-dependent phosphorylation of membrane of sarcolemma and sarcoplasmic reticulum of myocardium of rats during exercise. Proceedings of the Academy of Science of the Ukrainian S.S.R. B(1), 64-67; 1985.

Kalinski, M.I.; Zemtsova, I.I. Effect of exercise training on cAMP-dependent phosphorylation of protein of sarcoplasmatic reticulum of muscle. 4th Union symposium "Cyclic nucleotide"; Minsk; 1982.

Kalinski, M.I.; Zemtsova, I.I.; Kurski, M.D.; Osipenko, A.A. Influence of training and exercise on content of 3′, 5′-AMP and activity of enzymes of its metabolism in skeletal muscles of rats. Ukrainian Biochem. J. 52(6):611-613; 1980.

Kalinski, M.I.; Zemtsova, I.I.; Osipenko, A.A. Adenyl cyclase system of skeletal muscle and heart—intracellular mechanism of adaptation. Theory and practice of physical culture (Moscow). 5:45-47; 1990.

Kameyama, M.; Hescheler, J.; Mieskes, G.; Trautwein, W. The protein-specific phosphatase I antagonizes the beta-adrenergic increase of the cardiac Ca^{++} current. Pflugers Arch. 407:461-463; 1986.

Karczewski, P.; Bartel, S.; Haase, H.; Krause, E.G. Isoproterenol induces both cAMP- and calcium-dependent phosphorylation of phospholamban in canine heart in vivo. Biomed. Biochim. Acta. 46:433-439; 1987.

Katz, A.M. Interplay between inotropic and lusitropic effects of cyclic adenosine monophosphate on the myocardial cell. Circulation. 82(Suppl. 1): 1-7; 1990.

Katz, A.M.; Colvin, A.; Ashavaid, T. Phospholamban and calciductin. J. Mol. Cell. Cardiol. 15:795-799; 1983.

Katz, A.M.; Tada, M.; Kirchberger, M.A. Control of calcium transport in the myocardium by the cyclic AMP protein kinase system. Adv. Cyclic Nucl. Res. 5:453-472; 1975.

Kemp, B.E. cAMP-dependent protein kinase substrate specificity. Proc. Australian Biochem. Soc. 12:13; 1979.

Kim, H.W.; Kim, D.H.; Ikemoto, N.; Kranias, E.G. Lack of effects of calcium calmodulin-dependent phosphorylation on Ca^{++} release from cardiac sarcoplasmic reticulum. Biochim. Biophys. Acta. 903:333-340; 1987.

Kim, H.W.; Steenaart, N.A.E.; Ferguson, D.G.; Kranias, E.G. Functional reconstitution of the cardiac sarcoplasmic reticulum Ca^{++}-ATPase with phospholamban in phospholipid vesicles. J. Biol. Chem. 265:1702-1709; 1990.

Kirchberger, M.A.; Tada, M. Effects of adenosine 3′, 5′-monophosphate dependent protein kinase on sarcoplasmic reticulum isolated from cardiac and slow and fast contracting skeletal muscles. J. Biol. Chem. 251:725-729; 1976.

Kirchberger, M.A.; Tada, M.; Katz, A.M. Adenosine 3′, 5′-monophosphate-dependent protein kinase-catalyzed phosphorylation reaction and its relationship to calcium transport in cardiac sarcoplasmic reticulum. J. Biol. Chem. 249:6166-6173; 1974.

Korovkin, B.F. Cyclase system and activity of lysosomes enzymes in normal and pathologic conditions. Bulletin of Academy of Medical Science (Westn. Acad. Med. Nauk U.S.S.R.). 9:69-74; 1982.

Kovacs, R.J.; Nelson, M.T.; Simmerman, H.K.B.; Jones, L.R. Phospholamban forms Ca^{++}-selective channels in lipid bilayers. J. Biol. Chem. 264:18364-18368; 1988.

Kranias, E.G. Properties of phosphoprotein phosphatase activity associated with cardiac sarcoplasmic reticulum. Biophys. J. 45:2-5; 1984.

Kranias, E.G. Regulation of calcium transport by protein phosphatase activity associated with cardiac sarcoplasmic reticulum. J. Biol. Chem. 260:11006-11010; 1985.

Kranias, E.G.; Di Salvo, J. A phospholamban protein phosphatase activity associated with cardiac sarcoplasmic reticulum. J. Biol. Chem. 261:10029-10032; 1986.

Kraus, W.E.; Bernard, T.S.; Williams, R.S. Interaction between sustained contractive activity and beta-adrenergic receptors in regulation of gene expression in skeletal muscles. Am. J. Physiol. 256 (Cell Physiol. 25):C506-C514; 1989.

Krause, E.G. Transient changes in enzyme activities and phosphorylation of phospholamban during the cardiac cycle. J. Mol. Cell. Cardiol. 20:20; 1988.

Krause, E.G.; Bartel, S.; Beyerdorter, I.; Freier, W.; Gerber, K.; Obst, D. Transient changes in cyclic AMP and in the enzymic activity of protein kinase and phosphorylase during the cardiac cycle in the canine myocardium and the effect of propranolol. Mol. Cell. Biochem. 89:181-186; 1989.

Krebs, E.G.; Beavo, J.A. Phosphorylation-dephosphorylation of enzymes. Annu. Rev. Biochem. 48:923-959; 1979.

Kurski, M.D.; Osipenko, A.A.; Kalinski, M.I.; Kondratyuk, T.P. Some properties of 3′, 5′-AMP-dependent protein kinase of skeletal muscle during exercise to exhaustion. Biochem. (Moscow). 43(10):1776-1781; 1978.

Lamers, J.M.J. Calcium transport systems in cardiac sarcolemma and their regulation by second messengers cyclic AMP and calcium calmodulin. Gen. Physiol. Biophys. 4:143-154; 1985.

Lamers, J.M.J.; Stinis, J.T. Regulation of Ca^{++} transport across cardiac plasma membrane. Biochem. Soc. Trans. vol. 9; 1981.

Lamers, J.M.J.; Stinis, J.T. Inhibition of Ca^{++}-dependent protein kinase and Ca/Mg ATPase in cardiac sarcolemma by anticalmodulin drug calmidazolium. Cell. Calcium. 4:281-294; 1983.

Lamers, J.M.J.; Stinis, J.T.; Dejonge, H.R. On the role of cyclic AMP and Ca-calmodulin-dependent phosphorylation in the control of (Ca-Mg)ATPase of cardiac sarcolemma. FEBS Lett. 127:139-143; 1981.

Lamers, J.M.J.; Weeda, E. Methods for studying phosphorylation in cardiac membranes. In: Dhalla, N.S., ed. Methods in studying cardiac membranes. Boca Raton: CRC Press Inc.; 1984; chap. 13.

Laurence, L.B.; Hayes, J.S.; Mayer, S.E. Hormonally specific phosphorylation of cardiac troponin I and activation of glycogen phosphorylase. Nature. 280:78-80; 1979.

Lawrence, J.C., Jr.; Salgiver, W.J. Evidence that levels of malate dehydrogenase and fumarase are increased by cAMP in rat myotubes. Am. J. Physiol. 247 (Cell Physiol. 16):C33-C38; 1964.

Le Peuch, C.J.; Haiech, J.; Demaille, J.G. Concerted regulation of cardiac sarcoplasmic reticulum calcium transport by cyclic adenosine monophosphate-dependent and calcium-calmodulin-dependent phosphorylations. Biochem. 18:5160-5167; 1980.

Lincoln, T.M.; Corbin, J.D. On the role of the cAMP and cGMP dependent protein kinases in cell function. J. Cycl. Nucl. Res. 4:3-14; 1978.

Lindemann, J.P. Beta-adrenergic stimulation of sarcolemmal protein phosphorylation and slow responses in intact myocardium. J. Biol. Chem. 261:4860-4862; 1985.

Lindemann, J.P.; Jones, L.P.; Hathaway, D.R.; Henry, B.C.; Watanabe, A.M. Beta-adrenergic stimulation on phospholamban phosphorylation and Ca^{++}-ATPase activity in guinea pig ventricles. J. Biol. Chem. 258:464-471; 1983.

Lindemann, J.P.; Watanabe, A.M. Phosphorylation of phospholamban in intact myocardium. J. Biol. Chem. 260:4516-4525; 1985.

Litosch, I. Regulatory GTP-binding proteins: emerging concepts on their role in cell function. Life Sci. 41:251-258; 1987.

Malosheva, V.A.; Matlina, E.S. State of catecholamine synthesis during development of muscular fatigue. Bull. Exp. Biol. Med. (Moscow). 75:53-58; 1973.

Manalan, A.S.; Jones, L.R. Characterization of the intrinsic cAMP-dependent protein kinase activity and endogenous substrates in highly purified cardiac sarcolemmal vesicles. J. Biol. Chem. 257:10052-10062; 1982.

Marsden, C.D.; Meadows, J.C. The effect of adrenaline on the contraction of human muscle. J. Physiol. (London) 207:429-448; 1970.

Mazzeo, R.S. Catecholamine responses to acute and chronic exercise. Med. Sci. Sports Exerc. 23(7):839-845; 1991.

Mazzeo, R.S.; Grantham, P.A. Norepinephrine turnover in various tissues at rest and during exercise: evidence for a training effect. Metabolism. 38(5):479-483; 1989.

McDongall, L.; Jones, L.R.; Cohen, P. Identification of major protein phosphatases in mammalian cardiac muscle which dephosphorylate phospholamban. Eur. J. Biochem. 196:725-734; 1991.

Mery, P.F.; Brechler, V.; Pavoine, C.; Pecher, F.; Fischn eister, R. Glucagon stimulates the cardiac Ca current by activation of adenyl cyclase and inhibition of phosphodiesterase. Nature. 345:158-161; 1990.

Miller, W.C.; Gorski, J.; Oscai, L.B.; Palmer, W.K. Epinephrine activation of heparin nonreleaseable lipoprotein lipase in 3 skeletal muscle fiber types of the rat. Biochem. Biophys. Res. Commun. 164(2):615-619; 1989.

Miyakoda, G.; Yoshida, A.; Takisawa, H.; Nakamura, T. Involvement of phospholamban and/or 15-kDa protein phosphorylation in the increased contractile activity of the permeabilized rat heart cells. J. Mol. Cell. Cardiol. 20 (Suppl. 1):50; 1988.

Moir, A.J.; Cole, H.A.; Perry, S.V. The phosphorylation sites of troponin T from white skeletal muscle and the effect of interaction with troponin C on their phosphorylation by phosphorylase kinase. Biochem. J. 161:371-382; 1977.

Molla, A.; Capony, J.P.; Demaille, J.G. Cardiac sarcoplasmic reticulum calmodulin-binding proteins. Biochem. J. 226:859-865; 1985.

Moore, R.L.; Riedy, M.; Gollnick, P.D. Effect of training on beta-adrenoreceptor number in rat heart. J. Appl. Physiol.: Respir. Environ. Exerc. Physiol. 52(5):1133-1137; 1982.

Mundina, C.; Vittone, L.; Chiappo, G.; De Gingoiani, M.A. Phospholamban phosphorylation in the intact heart. J. Mol. Cell. Cardiol. 21 (Suppl. 2):184; 1989.

Murray, K.J.; Reeves, M.L.; England, P.J. Protein phosphorylation and compartments of cyclic AMP in the control of cardiac contraction. Mol. Cell. Biochem. 89:175-179; 1989.

Namm, D.H.; Mayer, S.E. Effect of epinephrine on cardiac cyclic 3′, 5′-AMP, phosphorylase kinase and phosphorylase. Mol. Pharmacol. 4:61-69; 1968.

Nawrath, H. Adrenoreceptor-mediated changes of excitation and contraction in isolated heart muscle preparation. J. Cardiovasc. Pharmacol. 14(3):1-10; 1989.

Noakes, T.D.; Higginson, L.; Opie, L.H. Physical training increases the ventricular fibrillation threshold of isolated rat hearts during normoxia, hypoxia and regional ischemia. Circulation. 67:24-30; 1983.

Nunoki, K.; Florio, V.; Catterall, W.A. Activation of purified calcium channels by stoichiometric protein phosphorylation. Proc. Natl. Acad. Sci. U.S.A. 86:6816-6820; 1989.

Opie, L.H. Review: role of cyclic nucleotides in heart metabolism. Cardiovasc. Res. 16:483-507; 1982.

Orlov, S.N., ed. Calmodulin. Moscow: VINITI AN USSR; 1987, p. 12-14.

Oscai, L.B.; Gorski, J.; Miller, W.C.; Palmer, W.K. Role of the alkaline TG lipase in regulating intramuscular TG content. Med. Sci. Sports Exerc. 30(6):539-544; 1988.

Osterrieder, W.; Brum, G.; Hescheler, J.; Trautwein, W.; Flockerzi, V. Injection of subunits of cyclic AMP-dependent protein kinase into cardiac myocytes modulates Ca current. Nature. 298:576-578; 1982.

Ostman, I.; Sjostrand, N.O. Cardiac noradrenaline turnover and urinary catecholamine excretion in trained and untrained rats during rest and exercise. Acta Physiol. Scand. 86(2):299-308; 1972.

Palmer, W.K. Effect of exercise on cardiac cyclic AMP. Med. Sci. Sports Exerc. 20(6):525-530; 1988.

Palmer, W.K.; Caruso, R.A.; Oscai, L.B. Cyclic AMP activation of a triglyceride lipase in broken cell preparations of rat heart. Arch. Biochem. Biophys. 249:255-262; 1986.

Palmer, W.K.; Doukas, S. Cyclic AMP phosphodiesterase activity in the hearts of trained rats. Can. J. Physiol. Pharmacol. 61:1017-1024; 1983.

Palmer, W.K.; Oscai, L.B.; Betchtel, P.J.; Fischer, G.A. Dibutyryl cAMP-induced increases in triacylglycerol lipase activity in developing L8 myotube cultures. Can. J. Physiol. Pharmacol. 68(6):689-693; 1990.

Palmer, W.K.; Studney, T.A.; Doukas, S. Exercise-induced increases in myocardial adenosine 3′, 5′-cyclic monophosphate and phosphodiesterase activity. Biochim. et Biophys. Acta. 672:114-122; 1981.

Palmer, W.K.; Studney, T.A.; Pikramenos, S. Tissue cyclic AMP following an acute bout of exercise. Med. Sci. Sport. 12(2):110; 1980 (abstract).

Perry, S. The regulation of contractile activity in muscle. Biochem. Soc. Trans. 7:593-617; 1979.

Philipson, K.D. "Calciductin" and voltage-sensitive calcium uptake. J. Mol. Cell. Cardiol. 15:867-868; 1983.

Powell, T.; Noble, D. Calcium movements during each heart beat. Mol. Cell. Biochem. 89:103-108; 1989.

Raeymaekers, L.; Hoffman, F.; Casteels, R. Cyclic GMP-dependent protein kinase phosphorylates phospholamban in isolated sarcoplasmic reticulum from cardiac and smooth muscle. Biochem. J. 252:269-273; 1988.

Rawis, R.L. G-proteins. Research unravels their role in cell communication. Chem. End. Neur. 65:26-39; 1987.

Reddy, Y.S. Phosphorylation of cardiac regulatory proteins by cyclic AMP-dependent protein kinase. Am. J. Physiol. 231:1330-1336; 1976.

Redman, C.M. Proteolipid involvement in human erythrocyte membrane function. Biochim. Biophys. Acta. 282:123-134; 1972.

Reeves, M.L.; England, P.J.; Murray, K.J. Compartments of cyclic AMP-dependent protein kinase in perfused guinea-pig hearts. Biochem. Soc. Trans. 15:955-956; 1987.

Resink, T.J.; Coetzee, G.A.; Gevers, W. Cardiac myofibrillar phosphorylation and adenosine triphosphate activity. S. Afr. Med. J. 56:897-906; 1979.

Reuter, H.M. Properties of two inward membrane currents in the heart. Annu. Rev. Physiol. 41:413-424; 1979.

Reuter, H.M. Calcium channel modulation by neurotransmitters, enzymes and drugs. Nature. 301:569-574; 1983.

Rinaldi, M.L.; Capony, J.P.; Demaille, J.G. The cyclic AMP-dependent modulation of cardiac sarcolemmal slow calcium channels. J. Mol. Cell. Cardiol. 14:279-289; 1982.

Rinaldi, M.L.; Peuch, C.J.; Demaille, J.G. The epinephrine-induced activation of the cardiac slow Ca^{++} channel is mediated by the cAMP-dependent phosphorylation of a calciductin, a 23,000 sarcolemmal protein. FEBS Lett. 129:277-281; 1981.

Romey, G.; Garcia, L.; Dimitriadov, V.; Pincon-Raymond, M.; Rieger, F.; Lazdunski, M. Ontogenesis and localization of Ca^{++} channels in mammalian skeletal muscle in culture and role in excitation-contraction coupling. Proc. Natl. Acad. Sci. U.S.A. 86:2933-2937; 1989.

Rubio, R.; Bailey, C.; Villar-Palasi, C. Effects of cyclic AMP-dependent protein kinase on cardiac actomyosin increase in Ca-sensitivity and possible phosphorylation of TN-I. J. Cycl. Nucl. Res. 1:143-150; 1975.

Schimid, A.; Renaud, G.F.; Lazdunski, M. Short term and long term effects of beta-adrenergic effectors and cyclic AMP on nitrendipine-sensitive voltage-dependent Ca channels of skeletal muscle. J. Biol. Chem. 260:13041-13046; 1985.

Schlender, K.K.; Hedary, M.G.; Thysseril, T.J. Dephosphorylation of cardiac myofibril C-protein by phosphatase I and protein phosphatase 2A. Biochim. Acta. 928:312-319; 1987.

Schulman, H.; Hanson, P.I. Multifunctional Ca/calmodulin-dependent protein kinase. Neurochem. Res. 18:65-77; 1993.

Schutzle, U.B.; Wakelam, M.J.O.; Pette, D. Prostaglandin and cyclic AMP stimulate creatine kinase synthesis but not fusion in cultured embryonic chick muscle cells. Biochim. Biophys. Acta. 805:204-210; 1984.

Schwartz, A.; Entman, M.L.; Kanihe, K.; Lane, L.K.; Van Winkle, W.B.; Bornet, E.P. The rate of calcium uptake into sarcoplasmic reticulum of cardiac muscle and skeletal muscle. Effects of cyclic AMP-dependent protein kinase and phosphorylase kinase. Biochim. Biophys. Acta. 426:57-72; 1976.

Schwoch, G.; Trinczek, B.; Bode, C. Localization of catalytic and regulator subunits of cyclic AMP-dependent protein kinases in mitochondria from various rat tissues. Biochem. J. 270(1):181-188; 1990.

Severson, D.L.; Drummond, G.I.; Sulakhe, P.V. Adenylate cyclase in skeletal muscle. J. Biol. Chem. 247:2949-2958; 1972.

Sheldon, A.; Kirby, C.R.; Booth, F.W. Cyclic AMP levels in fast- and slow-twitch muscle: response to exercise. Med. Sci. Sports Exerc. 24(5):360; 1992 (abstract).

Simmerman, H.K.B.; Collins, J.H.; Theibert, J.L.; Wegener, A.D.; Jones, L.R. Sequences analysis of phospholamban. Identification of phosphorylation sites and two major structural domains. J. Biol. Chem. 261:13333-13341; 1986.

Sperelakis, N., ed. Physiology and Pathophysiology of The Heart. Boston: The Haque Dordrecht Lancaster. p. 241-277; 1984.

Sperelakis, N.; Wahler, G.M. Regulation of Ca^{++} influx in myocardial cells by beta-adrenergic receptors, cyclic nucleotides and phosphorylation. Mol. Cell. Biochem. 82:19-28; 1988.

St. Lovis, P.J.; Sulakhe, P.V. Phosphorylation of cardiac sarcolemma by endogenous and exogenous protein kinase. Arch. Biochem. Biophys. 198:227-240; 1979.

Stewart, A.A.; Ingebritsen, T.S.; Manalan, A.; Klee, C.B.; Cohen, P. Discovery of a Ca^{++}- and calmodulin-dependent protein phosphatase. FEBS Lett. 137:80-84; 1982.

Stoclet, J.C.; Boulanger-Saunier, C.; Lassegue, B.; Lugnier, C. Cyclic nucleotides and calcium regulation in heart and smooth muscle cells. Ann. N.Y. Acad. Sci. 522:106-115; 1988.

Stull, J.T.; Buss, J.E. Phosphorylation of cardiac troponin by cyclic adenosine 3′, 5′-monophosphate-dependent protein kinase. J. Biol. Chem. 252:851-857; 1977.

Sulakhe, P.V.; Drummond, G.I. Protein kinase catalyzed phosphorylation of muscle sarcolemma. Arch. Biochem. Biophys. 961:448-455; 1974.

Sulakhe, P.V.; St. Lovis, P.J. Passive and active calcium fluxes across plasma membranes. Prog. Biophys. Mol. Biol. 35:135-195; 1980.

Suzuki, T.; Wang, J.H. Stimulation of bovine cardiac SR Ca^{++} pump and blocking of phospholamban phosphorylation by a phospholamban monoclonal antibody. J. Biol. Chem. 261:7018-7022; 1986.

Syska, H.J.; Wilkinson, M.; Grand, R.J.A.; Perry, S.V. The relationship between biological activity and primary structure of troponin I from white skeletal muscle of the rabbit. Arch. Biochem. Biophys. 198:227-240; 1979.

Tada, M.; Inui, M. Regulation of calcium transport by the ATPase phospholamban system. J. Mol. Cell. Cardiol. 15:565-577; 1983.

Tada, M.; Inui, M.; Jamada, M.; Kadoma, M.; Kusuya, T.; Abe, H.; Kakiuchi, S. Effects of phospholamban phosphorylation catalyzed by adenosine 3′, 5′-monophosphate and calmodulin-dependent protein kinases on calcium transport ATPase of cardiac sarcoplasmic reticulum. J. Mol. Cell. Cardiol. 15:335-346; 1983.

Tada, M.; Katz, A.M. Phosphorylation of the sarcoplasmic reticulum and sarcolemma. Annu. Rev. Physiol. 44:401-423; 1982.

Tada, M.; Kirchberger, M.A.; Repke, D.L.; Katz, A.M. The stimulation of calcium transport in cardiac sarcoplasmic reticulum by adenosine 3′, 5′-monophosphate-dependent protein kinase. J. Biol. Chem. 249:6174-6180; 1974.

Takeda, N.; Dominiak, D.; Turck, D.; Rupp, D.; Jacob, R. The influence of endurance training on mechanical catecholamine responsiveness, beta-adrenoreceptor density and myosin isoenzyme pattern of rat ventricular myocardium. Basic Res. Cardiol. 80:88-99; 1985.

Tate, C.A.; Taffet, G.E.; Hudson, E.K.; Blaylock, S.L.; McBride, R.P.; Michael, L.H. Enhanced calcium uptake of cardiac sarcoplasmic reticulum in exercise-trained old rats. Am. J. Physiol. 258 (Heart Circ. Physiol. 27):H431-H435; 1990.

Taylor, C.W. The role of G proteins in transmembrane signalling. Biochem. J. 272:1-13; 1990.

Tibbets, G.F.; Kashihara, H.; O'Reilly, K. Na^{+}-Ca^{2+} exchange in cardiac sarcolemma: modulation of Ca^{2+} affinity by exercise. Am. J. Physiol. 256 (Cell Physiol. 25):C638-C643; 1989.

Tsien, R.W. Cyclic AMP and contractile activity in heart. Adv. Cyclic Nucl. Res. 8:363-420; 1977.

Tuana, B.S.; Murphy, B.J.; Schwarzkopf, G. A calmodulin dependent protein kinase activity associated with rabbit heart sarcolemma. Mol. Cell. Biochem. 78:47-54; 1989.

Van Winkle, W.B.; Entman, M.L. Comparative aspects of cardiac and skeletal muscle sarcoplasmic reticulum. Life Sci. 25:1189-1200; 1979.

Varsanyi, M.; Tolle, H.C.; Heilmeyer, L.M.G.; Dawson, R.M.C.; Irvine, F.R. Activation of sarcoplasmic reticular Ca transport ATPase by phosphorylation of an associated phosphatidylinositol. EMBO J. 2:1543-1548; 1983.

Velema, J.; Noordam, P.C.; Zaagama, J. Comparison of cyclic AMP-dependent phosphorylation of sarcolemma and sarcoplasmic reticulum from rat ventricle muscle. Int. J. Biochem. 15:675-684; 1983.

Vetter, R.; Haase, H.; Will, H. Potentiating effect of calmodulin and catalytic subunit of cyclic AMP-dependent Ca^{++} transport by cardiac sarcolemma. FEBS Lett. 148:326-333; 1982.

Vihko, V.; Hirsimaki, Y.; Adjadman, A.O. b-Glucuronidase activity in trained red and white skeletal muscle of mice. Eur. J. Appl. Physiol. 39(4):255-261; 1978.

Vihko, V.; Salminen, A.; Rantamaki, J. Acid hydrolase activity in red and white skeletal muscle of mice during a two week period following exhausting exercise. Pflugers Arch. 378(2):99-106; 1978.

Vorobetz, Z.P.; Kocherga, V.I.; Nesterenko, N.V.; Kurchenko, L.K. Ca^{++}-phospholipid-dependent phosphorylation of the myocardial sarcolemma vesicular preparations. Biochem. (Moscow). 53:1327-1333; 1988.

Vorotnicov, A.V.; Risnik, V.V.; Gusev, V.B. Phosphorylation of the heart and skeletal muscle troponin by Ca^{++}-phospholipid-dependent protein kinase. Biochem. (Moscow). 53(1):31-40; 1988.

Wahler, G.M.; Rush, N.J.; Sperelakis, N. Inhibition of whole-cell calcium channel currents by 8-bromo-cGMP in voltage clamped embryonic chick heart myocytes. Physiologist. 30:178; 1987.

Walaas, O.; Walaas, E.; Gronnerod, O. Hormonal regulation of cyclic AMP-dependent protein kinase of rat diaphragm by epinephrine and insulin. Eur. J. Biochem. 40:465-477; 1973.

Walaas, S.I.; Horn, R.S.; Nairn, A.C.; Walaas, O.; Adler, A. Skeletal muscle sarcolemma proteins as targets for adenosine 3′, 5′-monophosphate-dependent and calcium-dependent protein kinases. Arch. Biochem. Biophys. 262:245-258; 1988.

Walsh, M.P.; Cavadore, J.C.; Vallet, B.; Demaille, J.G. Calmodulin-dependent myosin light chain kinases from cardiac and smooth muscle: a comparative study. Can. J. Biochem. 58:299-308; 1980.

Watanabe, A.M.; Lindemann, J.P.; Jones, L.R.; Besch, H.R.J.; Bailey, J.C. Biochemical mechanisms mediating neural control of the heart. In: Abboud et al., eds. Disturbances in neurogenic control of circulation. Baltimore: Waverly; 1981: p. 189-203.

Watson, P.A.; Haneda, T.; Morgan, H.E. Effect of higher aortic pressure of ribosome formation and cAMP content in rat heart. Am. J. Physiol. 256 (Cell Physiol. 25):C1257-C1261; 1989.

Wegener, A.; Simmerman, H.K.B.; Lindemann, J.P.; Jones, L.R. Phospholamban phosphorylation in intact ventricles. Phosphorylation of serine 16 and threonine 17 in response to beta-adrenergic stimulation. J. Biol. Chem. 264:11468-11474; 1989.

Wegener, A.B.; Jones, L.R. Phosphorylation-induced mobility shift in phospholamban in sodium dodecyl sulphate-polyacrylamide gels. J. Biol. Chem. 259:1831-1834; 1984.

Werle, E.O.; Strobel, G.; Weicker, H. Decrease in rat cardiac beta 1- and beta 2-adrenoreceptors by training and endurance exercise. Life Sci. 46(1):9-17; 1990.

Westwood, S.A.; Perry, S.V. The effect of adrenaline on the phosphorylation of the P light chain of myosin and troponin I in the perfused rabbit heart. Biochem. J. 197:185-193; 1981.

Will, H. Zur rolle von Proteinphosphorylierungen in der Zellregulation. Deutsch Gesundheistw. 38:201-205; 1983.

Williams, R.S.; Caron, M.G.; Daniel, K. Skeletal muscle beta-adrenergic receptors: variations due to fiber type and training. Am. J. Physiol. 246 (Endocrinol. Metab. 9):E160-E167; 1984.

Winder, W.W.; Arogyasami, J.; Barton, R.J.; Elayan, I.M.; Vehrs, P.H. Muscle malonyl-CoA decreases during exercise. J. Appl. Physiol. 67(6):2230-2233; 1989.

Winder, W.W.; Duan, C. Control of fructose 2,6-diphosphate in muscle of exercising fasted rats. Am. J. Physiol. 262 (Endocrinol. Metab. 25):E919-E924; 1992.

Winder, W.W.; Fisher, S.R.; Gygi, S.P.; Mitchell, J.A.; Ojuka, E.; Weidman, D.A. Divergence of muscle and liver fructose 2,6-diphosphate in fasted exercising rats. Am. J. Physiol. 260 (Endocrinol. Metab. 23):E756-E761; 1991.

Winegrad, S.; McClellan, G.; Horowits, R.; Ticker, M.; Lin, L.; Weisberg, A. Regulation of cardiac contractile proteins by phosphorylation. Fed. Proc. 42:39-44; 1983.

Winegrad, S.; McClellan, G.; Tucker, M.; Lin, L.E. Cyclic AMP regulation of myosin isozymes in mammalian cardiac muscle. J. Gen. Physiol. 81:749-765; 1983.

Wollenberger, A.; Will, H. Protein kinase catalyzed membrane phosphorylation and its possible relationship to the role of calcium in the adrenergic regulation of cardiac contraction. Life Sci. 22:1159-1178; 1978.

Wray, H.L.; Gray, R.R. Cyclic AMP-stimulation of membrane phosphorylation and Ca^{++}-activated ATPase in cardiac sarcoplasmic reticulum. Biochim. Biophys. Acta. 461:442-459; 1977.

Wuytack, F.; Schutter, G.D.; Casteels, R. The effect of calmodulin on the calcium ion transport and (Ca^{++}, Mg^{++})-dependent ATPase in microsomal fractions of smooth muscle. Biochem. J. 190:827-831; 1980.

Wyatt, H.L.; Church, L.; Rabinowitz, B.; Tyberg, J.V.; Parmley, W.W. Enhanced cardiac response to catecholamines in physically trained cats. Am. J. Physiol. 234 (Heart Circ. Physiol. 3)H608-H613; 1978.

Xenophontos, X.P.; Watson, P.A.; Chua, B.H.A.; Haneda, T.; Morgan, H.E. Increased cyclic AMP content accelerates protein synthesis in rat heart. Circ. Res. 65:647-656; 1989.

Yeaman, S.J.; Cohen, P. The hormonal control of activity of skeletal muscle phosphorylase kinase. Phosphorylation of the enzyme at two sites in vivo in response to adrenaline. Eur. J. Biochem. 51:93-104; 1975.

Zeman, R.J.; Ludemann, R.; Easton, R.G.; Ettinger, J.D. Slow to fast alteration in skeletal muscle fibers caused by clenbuterol, a beta-2-receptor agonist. Am. J. Physiol. 254 (Endocrinol. Metab. 17):E726-E732; 1988.

Zemtsova, I.I.; Kalinski, M.I.; Kurski, M.D.; Osipenko, A.A. Influence of thermostable protein inhibitor on activity of 3′, 5′-cAMP-dependent protein kinase of skeletal muscle of rats during endurance training. Proceedings of the Academy of Science of Ukrainian S.S.R. B(2):72-75; 1981.

Ziegelhoffer, A.; Anand-Srivastava, M.B.; Khandelwal, R.L.; Dhalla, N.S. Activation of heart sarcolemmal Ca^{++}/Mg^{++}-ATPase by cyclic AMP-dependent protein kinase. Biochem. Biophys. Res. Commun. 89:1073-1081; 1979.